SOTHIS PRESS

Sothis Press aims to completely reimagine the public domain book with an emphasis on universal access to captivatingly designed print editions. We believe that technology will enable every single book ever published to be perpetually accessible in beautiful print form. Sothis is beauty in ubiquity.

www.sothis.press
1 Waterhouse Square, London, ECIN 2ST

A Course in Descriptive Geometry and Photogrammetry for the Mathematical Laboratory

E. L. Ince

SOTHIS PRESS

ISBN 978-1-0365-1867-7

A Course in Descriptive Geometry and Photogrammetry for the Mathematical Laboratory is in the public domain.

This edition is published by Sothis Press © 2023. The company is registered in England and Wales. The Company Number is 14900389.

Edinburgh Mathematical Tracts

No. 1

DESCRIPTIVE GEOMETRY AND PHOTOGRAMMETRY

A COURSE IN DESCRIPTIVE GEOMETRY AND PHOTOGRAMMETRY

FOR THE

MATHEMATICAL LABORATORY

BY

E. LINDSAY INCE, M.A., B.Sc.
BAXTER RESEARCH SCHOLAR IN THE UNIVERSITY OF EDINBURGH

LONDON
G. BELL & SONS, LIMITED
YORK HOUSE, PORTUGAL STREET
1915

PREFACE

ONE of the most desirable attainments of the mathematician is the ability to form a mental picture of three dimensional systems with ease and perspicuity, an ability which cannot better be developed than by a course of Descriptive Geometry.

It is a singular and regrettable fact that in most British Universities, Descriptive Geometry has hitherto been omitted from the regular mathematical curriculum and studied only in the technical classes. The tendency of recent years, however, is towards a recognition of its unique value, both from the educational and from the practical point of view.

The present tract embodies the course which is given to the non-technical students in the Mathematical Laboratory of the University of Edinburgh. In its compilation I have consulted from time to time the works of Catalan, Antomari, and Gino Loria, and, above all, the *Géométrie Descriptive* of Monge, which has been a never-failing source of inspiration.

I cannot bring this Preface to a conclusion without expressing my sense of deep obligation to Professor Whittaker for his invaluable suggestions and criticism, both during the compilation of the tract and during its passage through the press.

E. L. I.

THE MATHEMATICAL LABORATORY,
UNIVERSITY OF EDINBURGH,
23rd June 1915.

CONTENTS

CHAPTER I

Introduction

CHAPTER II

The Straight Line and Plane in Orthogonal Projection

CHAPTER III

Curved Surfaces and Space-Curves in Orthogonal Projection

CHAPTER IV

Perspective

CHAPTER V

Photogrammetry

CHAPTER I

INTRODUCTION

1. The Purpose of Descriptive Geometry.—There is one great difficulty that every student of solid geometry sooner or later encounters, viz. that of evolving some plane representation of the lines, surfaces, etc., with which he is dealing, which is not a mere diagrammatic illustration invented for the sole purpose of assisting the eye and brain, and incapable of representing the true metrical relation of one part of the figure to another. In plane geometry, such a difficulty never arises, for all the points, straight lines, and curves which are dealt with can be represented in their true relative positions and all drawn to one convenient scale. Thus the diagram produced is not only an aid to the eye and brain as showing the true nature of the problem dealt with, but is also a convenient and accurate instrument by means of which the properties of the figures concerned may be investigated. The science of *Descriptive Geometry* * may be said to be an attempt to extend these advantages that plane geometry is seen to possess to the geometry of three dimensions also, or, in other words, to solve by the methods of plane geometry, problems connected with the geometry of space. It has, therefore, some considerable importance not only from the theoretical but also from the technical point of view in its applications to the applied sciences of engineering and architecture.

2. The Methods of Descriptive Geometry.—As would be expected, the accurate representation on a plane of a figure in three dimensions depends on some method of projection. But since a point in space requires three co-ordinates to determine

* Fr. *Géométrie descriptive*, Ger. *Darstellende Geometrie.*

its position and a point in a plane is completely fixed if two co-ordinates associated with it are known, no single projection will, unaided, give an accurate representation of a figure in space. In other words, though the projection of a figure in space on a plane is unique, the converse is not true, and some more or less artificial device must be evolved by which the plane representation uniquely corresponds to the relative figure in space. Several devices of this nature are known, of greater or less merit, but only three will be given in this tract. The first and by far the most important of these is the *method of doubly-orthogonal projection*, or *Monge's Method* (1798), which on account of its directness no less than of its historical interest naturally occupies the greater number of the succeeding pages. The other two are respectively the method of *contour lines*, which has so wide an application in the art of map-making, and the method of *central projection* or *perspective* with which the names of Brook Taylor and Lambert, and, more recently, of Fiedler, are closely associated. In the following sections these three methods of descriptive representation will be separately explained, special reference being made in each case to the fundamental elements—the point, the straight line, and the plane.

3. Monge's Method.—Although, as has been remarked, a point in space is not uniquely determined if its projection on one single plane is known, yet it is uniquely determined if its projections are known on two planes which are not parallel to one another. This is the foundation of Monge's Method, in which case the two planes considered are at right angles to one another. For convenience one of these planes is taken to be horizontal, and is called the *horizontal plane of projection*,* the other being naturally known as the *vertical plane of projection*.† The intersection of the two planes is a horizontal line called the *ground line*.‡

Suppose now that it is desired to represent by Monge's Method the straight line AB in space (fig. 1). Project it orthogonally on to the horizontal and vertical planes, the projections being ab and a_1b_1 respectively. Now rotate the

* Fr. *plan horizontal*, Ger. *Grundrissebene*.

† Fr. *plan vertical*, Ger. *Aufrissebene*.

‡ Fr. *ligne de terre*, Ger. *Grundlinie* or *Projektionsachse*.

vertical plane of projection about the ground line XY, as if the latter formed a hinge, until the two planes are co-planar, the vertical plane now coinciding with the continuation backwards past the ground line of the horizontal plane. The vertical projection a_1b_1 of the straight line AB now takes up a position $a'b'$ in the same plane as the horizontal projection ab.

The straight line AB in space is now represented not by its projections on two perpendicular planes but by a *plane* diagram consisting of the ground line XY and the two straight lines ab

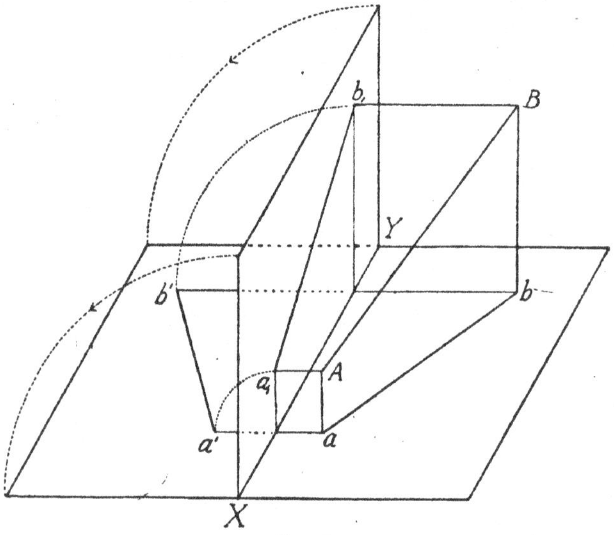

FIG. 1.

and $a'b'$, which represent the horizontal and vertical projections respectively (fig. 2).

This operation of turning a plane into coincidence with another plane is of frequent application in the processes of descriptive geometry, and is known under the name of *rabatting.** Several illustrations of the power and scope of this method will be seen in the next chapter.

It is always advisable, if possible, so to choose the planes of projection so that the projected figure is above the horizontal

* Fr. *rabattre*, Ger. *umdrehen*.

plane and in front of the vertical plane, that is to say, occupies a position similar to that of the line AB in fig. 1. Such being the case, the horizontal projection is entirely below and the vertical projection entirely above the ground line in the rabatted figure, so that the two projections do not encroach upon each other. It may happen, however, that in certain cases the horizontal projection of a figure in space actually crosses the ground line and intrudes into the region occupied by the rabatted vertical projection, and *vice versa*. Such a case would arise, for instance, if the line AB (fig. 1) were produced backwards so as ultimately to pass through the vertical plane

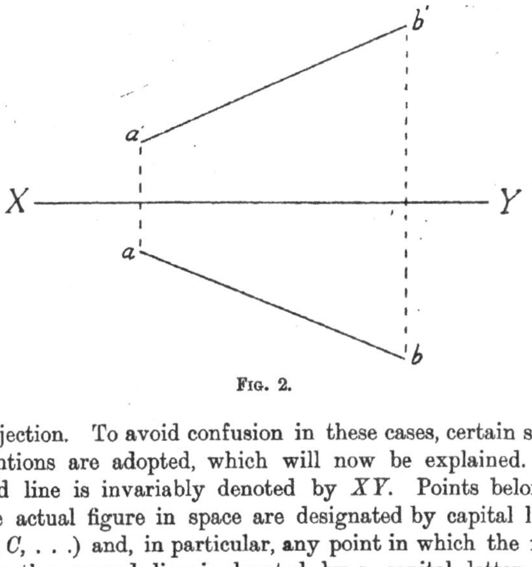

FIG. 2.

of projection. To avoid confusion in these cases, certain simple conventions are adopted, which will now be explained. The ground line is invariably denoted by XY. Points belonging to the actual figure in space are designated by capital letters (A, B, C, . . .) and, in particular, any point in which the figure crosses the ground line is denoted by a capital letter. The corresponding points of the horizontal projection are denoted by the corresponding small letters (a, b, c, . . .), and those of the rabatted vertical projection by the corresponding small accented letters (a', b', c', . . .). Thus a and a' are the horizontal and vertical projections respectively of the point A of the projected figure, and A is now known as "the point (a, a')." All lines belonging properly to the horizontal projection which appear above the ground line are *dotted*, and likewise those belonging to the vertical projection which appear below the ground line (*cf.* Chap. II., fig. 10).

4. Fundamental Properties of the Projections.—It is easily seen that the horizontal projection a and the vertical projection a' must lie on a line perpendicular to the ground line. For since Aa and Aa_1 (fig. 1) are perpendicular to the horizontal and vertical planes of projection respectively, the plane aAa' which contains them must also be perpendicular to these two planes, and hence to their line of intersection XY. It follows, therefore, that the line aa', in which the plane aAa' cuts the horizontal plane, stands at right angles to the ground line XY. Thus a complete specification of the point (a, a') involves the knowledge of the perpendicular distances of a and a' below and above the ground line respectively, and the distance of the foot of their common perpendicular on the ground line from a chosen base-point on XY. Thus to determine a point in space by means of its projections *three* data are required, corresponding to the three co-ordinates by which the position of a point in space may be specified.

Though it is usual and convenient to define a straight line by its projections on the two reference planes, the line may equally well be determined by the two points in which it intersects the horizontal and vertical planes of projection. These two points are called respectively the horizontal and vertical *traces* * of the line in question. It follows from the definition that the horizontal trace is its own horizontal projection, the corresponding vertical projection being the foot of the perpendicular drawn on to the ground line. Likewise the horizontal projection of the vertical trace is obtained by dropping a perpendicular from it on to the ground line.

An unlimited plane in space cannot be determined by its projections on the reference planes, for these projections would in general be simply the reference planes themselves. The most convenient way of representing a plane is by its lines of intersection with the reference planes, for it is obvious that if these lines are known, the plane is uniquely determined, except in the indeterminate case in which it passes through the ground line.

* Fr. *trace*, Ger. *Spur*.

By analogy with the case of a straight line in space, the name of *traces* is given to these lines of intersection, for they contain the corresponding traces of all straight lines drawn in the plane itself. Furthermore, since the plane considered cuts the ground line in one unique point, the horizontal and vertical traces of the plane must meet on the ground line. If aB and Bc' be the horizontal and vertical traces respectively, the plane corresponding may be designated as "the plane aBc'." From what has been said several deductions are immediately obvious, viz.:—

I. If the plane is parallel to one of the planes of projection, it has no trace on the latter, or, as we may say, the corresponding trace is at infinity. Its trace on the other plane is parallel to the ground line.

II. If the plane is parallel to the ground line, but cuts both of the planes of the projection, then its traces are both parallel to the ground line.

III. If the plane is perpendicular to the ground line, the two traces coincide in a single straight line perpendicular to the ground line.

5. The Method of Contours.—A point in space is determined if its projection on a fixed horizontal plane and its distance above or below that plane are known. Similarly, a straight line in space is completely determined if its horizontal projection and the heights of any two points on it above or below the plane of projection are known. On these simple principles depends the theory of the method of contours. A solid body, being made up of points, is completely represented if a number of points on it are known, sufficient to fix its position in space, and thus a map or diagram of the body can be formed, in the following manner:—A convenient scale being chosen, the *level* * or distance of each required point from the horizontal plane is written beside its horizontal projection, and this distance is regarded as positive or negative according as the projected point is above or below the horizontal plane. A horizontal line has always the same level, and this is denoted by writing that level close to the projection of the line itself, *e.g.* the straight line AB (fig. 3) has the level 4·2. A straight line inclined to the horizontal is best denoted by marking on it those points

* Fr. *cote*, Ger. *Kote*.

whose levels are integral multiples (1, 2, 3,. . .) of the scale unit, as is seen in the case of the line *CD*. The point of zero level on the line is, of course, its trace on the horizontal plane. A plane is uniquely represented by a line of steepest slope, that is to say, any line in the plane itself which cuts at right angles the horizontal lines of the plane. To distinguish it from any other line, it is denoted by a double line such as *EF*. The apparent or horizontal distance between two consecutive points whose levels are integral multiples of the scale unit is called the *interval*. Thus the smaller the interval, the greater is the

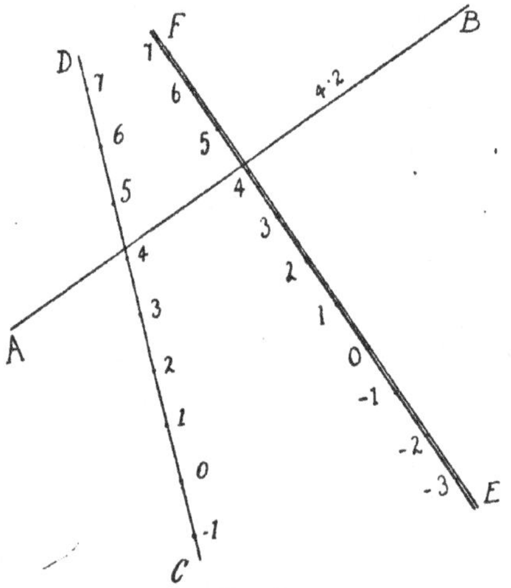

FIG. 3.

inclination of the corresponding line or plane to the horizontal. The intervals of a straight line or plane are obviously all equal to one another.

6. The Method of Perspective.—The two methods which have so far been dealt with are distinguished by the fact that they are both orthogonal projections, so that the projection of a point is simply the foot of the perpendicular drawn from it on to the plane of projection. Thus the lines joining the projections of a number of points in space to the corresponding points themselves are parallel and meet only at an infinite distance from the plane of projection, or, in other words, the *centre of projection* is at infinity. In the method of perspective or central projection, on the other hand, the centre of projection, which is then generally called the *station point* or *point of sight*,* is at a finite distance from the projected object, and the projection of a given point is the trace on a single plane of projection of the straight line or *ray* drawn through the centre of projection and the given point.

Suppose now that it is required to represent the straight line AB in perspective with a given station point or centre C on a given plane of projection or *picture plane*,† which may

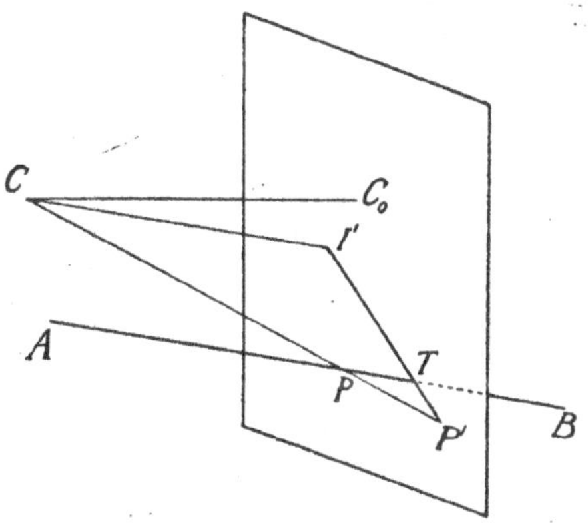

FIG. 4.

conveniently be taken to be vertical. In order that the position of C may be determined, the perpendicular CC_0 is drawn on to the plane of projection, so that if the foot C_0 of this perpendicular (which is called the *centre of vision* ‡), and its length are known, then C is determined. Now the projection of AB is formed of the projections of all the points situated along it, and is therefore the straight line $I'T$ in which the plane through the line AB itself and the centre of projection C intersects the plane of projection. This is not, however, a unique representation, in the sense that it represents the line

* Known also as the *centre of perspective*. Fr. *point de vue*, Ger. *Projektionszentrum*, *Augenpunkt*.

† Fr. *plan du tableau*, Ger. *Bildebene*.

‡ Fr. *point principal*, Ger. *Hauptpunkt*.

AB and no other, for it is equally the projection of any straight line whatsoever in the plane through C and AB. It is therefore necessary to evolve some method by which the representation may be rendered unique. For this purpose two points are chosen on the projected line $I'T$ which not only define the projection itself, but are such that the line in space is also determined by them. Suppose, therefore, that T is the *trace* of the line AB, and is therefore a point also on the projected line. If T is assigned, the line AB has now only one degree of freedom remaining, and if this be destroyed, AB is fixed in space. Suppose, therefore, that through the centre of projection C a line CI', whose trace is I', be drawn parallel to AB, so

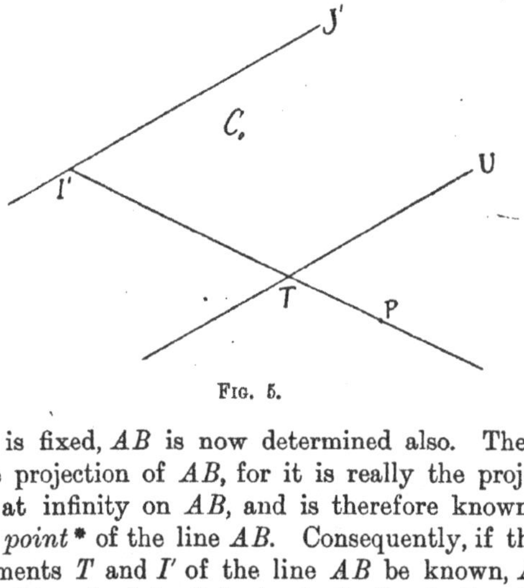

FIG. 5.

that if CI' is fixed, AB is now determined also. The point I' lies on the projection of AB, for it is really the projection of the point at infinity on AB, and is therefore known as the *vanishing point** of the line AB. Consequently, if the fundamental elements T and I' of the line AB be known, AB itself is uniquely determined, for it is simply the line in the plane $CI'T$ which passes through T and is parallel to CI', and is known as "the line (TI')." It is easily seen that for all straight lines having the same trace T the smaller the distance TI', which is known as the *interval* of the line, the more acute the angle which the line AB makes with CT, and *vice versa*.

If P be a point on AB and P' its projection, then P is uniquely determined if the points T, I', and P' (which are

* Fr. *point de fuite*, Ger. *Fluchtpunkt*.

evidently collinear) are known. This provides a means of representing a point in space, for it is completely determined if its projection P' and the trace T and the vanishing point I' of any straight line passing through it (but not through C) are known, so that the point P may be described as "the point (TI', P')." The line (TI') may be chosen at random.

A plane in space may be represented in a manner similar to the straight line, as it is uniquely determined by its *trace* TU and its *vanishing line* $I'J'$, the latter being the trace of a plane through C parallel to the given plane. Thus if a line lies in a given plane, its trace will lie on the trace of the plane and its vanishing point on the corresponding vanishing line.

From what has been said, the following conclusions are easily drawn:—

I. Parallel straight lines have the same vanishing point, and parallel planes have the same vanishing line.

II. The centre of vision is the vanishing point of all straight lines perpendicular to the picture plane.

III. If two straight lines intersect, the line joining their traces is parallel to the line joining their vanishing points.

IV. Points, straight lines, etc., situated in a plane through C parallel to the plane of projection all project to infinity. This plane is known as the *vanishing plane.*

V. If two planes intersect, the trace and vanishing point of the line of intersection are respectively the intersections of the traces and of the vanishing lines of the given planes.

7. Laboratory Methods.—As the purpose of descriptive geometry is to evolve methods by which solid figures may be represented with their component elements in their true metrical relations to one another, it follows that if the full benefits of these methods are to be attained, considerable precision in working must be aimed at. Diagrams should therefore be carefully drawn, on as large a scale as is practicable, and all unnecessary detail in construction suppressed. The materials required are few, and for work in the laboratory the following may be recommended:—Drawing board (imperial size) and T-square, divided ruler, set square, half set of compasses (6-inch), dividers, protractor and pricker. A supply of good cartridge paper, a moderately hard pencil, and drawing-pins are also required. The compasses should have needle points, and may with advantage be bow headed; they should be jointed in order that the limbs may stand at right angles to the paper, as otherwise large holes are liable tó be formed. The pricker is simply a needle mounted in a wooden handle, and should always be used instead of the pencil for marking points on the diagram. Measured distances should be set off from the ruler-edge with the pricker; other distances are transferred by the dividers. The pencil should have a sharp, chisel-shaped point, which will be found to give a finer straight line than a conical point and to be more durable; the compass pencil, however, must have a conical point.

In laboratory work, the first step in the construction is invariably to draw by means of the T-square a horizontal line right across the centre of the paper, to serve as an axis of reference and, if necessary, also as the ground line. Other horizontal lines may be drawn subsequently, as required, by the T-square. A vertical line is then drawn, near the centre of the paper, accurately at right angles to the horizontal line. This should be done with dividers, the method of Euclid I. 11 being adopted. This vertical line serves as the second axis of reference, and other vertical lines can be drawn as required by parallelism with this; the commonest method of drawing a line parallel to a given line is by sliding a set square worked against a fixed edge. In the subsequent construction arcs of circles should always be used in preference to straight lines, for a circle can be drawn more accurately than a straight line can be copied. Dividers should always be used in preference to compasses, except when the arc to be constructed is part of the final result, since the circle scratched out by the sharp point of the divider is more clearly defined than the pencil mark of the compasses. Dividers should be set to the correct radius by closing them from a larger, rather than by opening them from a smaller, radius.

Whenever possible, the results of the work should be verified by some means independent of the previous construction. Thus in Monge's Method, whenever a plane has to be found, its horizontal and vertical traces should be investigated separately, and then it should be ascertained whether or not they meet on the ground line. If they do not, the flaw in the construction should be rectified before the next step is proceeded with.

It may occasionally happen that during the course of the work practical difficulties arise owing to two lines being so nearly parallel to one another that their point of intersection is inaccessible. The method of treatment in such a case is best illustrated by means of an example. Suppose, therefore, that it is required to draw a line through a given point P and the inaccessible intersection of two given straight lines RS and TU. Through P draw PX and PY to cut RS and TU in R and T respectively. Let XYZ be any transversal, not passing

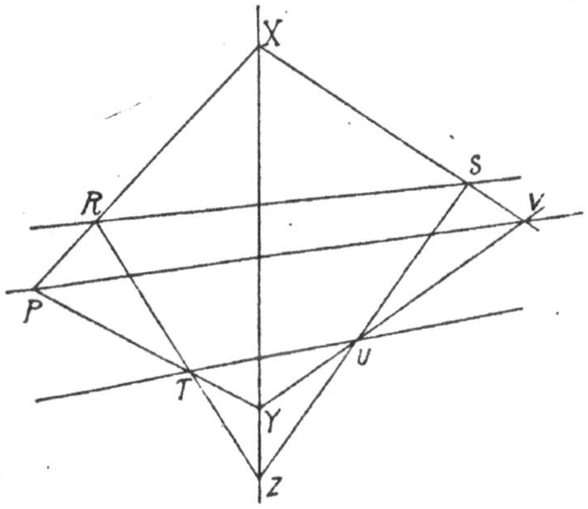

FIG. 6.

through P, cutting PX in X, PY in Y, and RT produced in Z. Through Z draw ZUS to cut TU and RS in U and S respectively, and join XS and YU, producing them until they meet in V. Then PV will pass through the intersection of RS and TU, for the triangles PRT and VSU are so situated that the intersections of their corresponding sides are collinear, and hence, by Desargues' Theorem, the lines RS, TU, and PV joining corresponding vertices are concurrent.

In dealing with plane curves, or with the projections of twisted curves in space, the problem very frequently arises of drawing a tangent line to the curve from a given point on or outside it. This construction may very conveniently be carried out in one or other of the following ways:—

Suppose, first of all, that the given point p lies on the given curve. With p as centre and any convenient radius describe a circle, and draw several radii pb_1, pb_2 . . . of this circle to cut the curve in a_1, a_2, . . . respectively. On the lines pb_1, pb_2, . . . mark off distances pc_1, pc_2, . . . equal to and in the same direction as a_1b_1, a_2b_2, . . . and let the curve traced out by the points c_1, c_2, . . . intersect the circle in q. Then pq will be the required tangent line (fig. 7).

Next, suppose that the given point p does not lie on the curve itself. Through p draw secants pa_1b_1, pa_2b_2, . . . cutting the curve in a_1 and b_1, a_2 and b_2. . . . Bisect a_1b_1 in c_1,

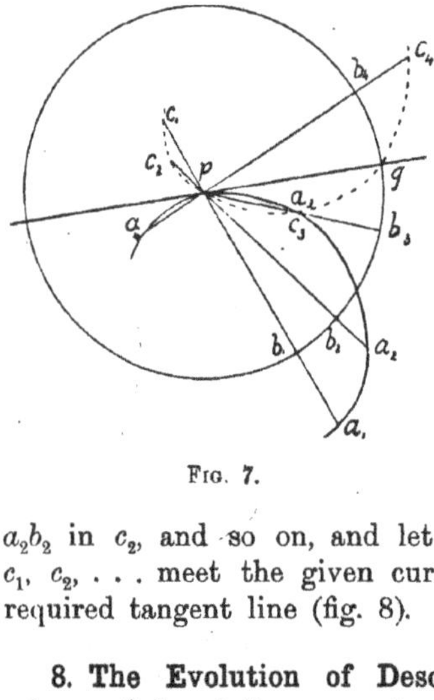

Fig. 7.

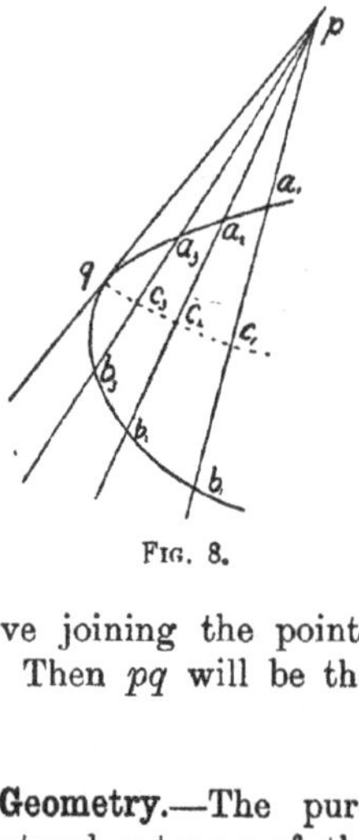

Fig. 8.

a_2b_2 in c_2, and so on, and let the curve joining the points c_1, c_2, . . . meet the given curve in q. Then pq will be the required tangent line (fig. 8).

8. The Evolution of Descriptive Geometry.—The pure science of descriptive geometry is the natural outcome of the study of those problems which the applied science of architecture has from time immemorial brought to the attention of mankind. The plans and elevations which were the forerunners of the horizontal and vertical projections of the descriptive geometry of Monge were certainly employed by the Egyptians and the Greeks, and Vitruvius, the renowned architect of the time of Augustus, makes explicit use in his *Architectura* of the terms "ichnographia" and "orthographia" to signify respectively the plan and elevation of a building. These diagrams, however, were regarded solely as convenient and accurate representations of figures in space, and nothing was known of their application to any but the very simplest of geometrical constructions. In short, this invaluable method of representing solid bodies was known and employed, but the possibility of adapting it to the discussion of the geometrical properties of the body in question appears to have remained undeveloped. It would seem that the art of Stereotomy, or wood and stone carving, to which so great attention was devoted in the Middle Ages, first revealed this possibility and indicated certain empirical rules by means of which geometrical constructions could be carried out. In the *Traité de l'Architecture* of Philibert de l'Orme many examples of constructions are given, by which plane sections of solid figures and the true positions of points on a solid body may be obtained. The problems of the development of a solid and of the effect of rotating the figure considered were also attempted, but no verification of the constructions given was attempted, and in very many cases the results were of extremely limited applicability.

From the theoretical point of view, the first name which falls to be mentioned is that of Desargues (1593–1662), who might indeed be regarded as the founder of modern descriptive geometry. His work was known to the world through his pupil and commentator Bosse. Unfortunately, he did not use his method as a foundation on which to build a connected science, but satisfied himself with adapting it to the solution of disconnected problems arising in practice. This great step was not taken until a century later, when comes the name of Frézier (1682–1773), whose work *La théorie et la pratique de la coupe des pierres et des bois . . . ou traité de stéréotomie à l'usage de l'architecture* is a classic among works on geometry, placing, as it did, the methods of descriptive geometry on a firm and connected scientific basis. In it plane curves and curved surfaces are dealt with; by the use of an auxiliary plane the sections of solids with one another are obtained, and methods of solving trihedral angles are investigated. These are, however, treated not as isolated problems but as examples which illustrate a general theory.

In Gaspard Monge (1746–1818), descriptive geometry reached

its height, for it was he who finally raised what was formerly merely a tool in the hand of the practical geometer to the dignity of an independent science, and to him also is the name "descriptive geometry" due. Of humble birth, he encountered great difficulties in his education, but his outstanding genius overcame all obstacles and found recognition in his appointment to a chair in 1768. At the time of the Revolution, Monge showed himself to be an eager supporter of the new régime, and his devotion was rewarded by several valuable appointments under the government. In 1798 he was made Professor at the École Polytechnique in Paris. There he gave a course of lectures on descriptive geometry, the material of which was collected and published in 1800 as the classical "géométrie descriptive" on which nearly all later works have been based. At the restoration Monge was deprived of his offices, and did not long survive his degradation. In Monge, descriptive geometry not only attained the dignity of a pure science, but showed itself capable of giving elegant solutions to the every-day problems of geodesy and topography. So worthy a place did he claim for it in the annals of mathematical science that his lectures drew from Lacroix the admiring words: "Tout ce que je fais par le calcul, je pourrais l'exécuter avec la règle et le compas, mais il ne m'est pas permis de vous révéler ces sécrets."

Concurrently with that method of descriptive geometry which culminated in the work of Monge arose and developed the method of perspective. Hipparchus (150 B.C.) was the inventor of stereographic projection, and this and other geographical projections were known to Ptolemy (135 A.D.). From the Middle Ages to the end of the seventeenth century the study of perspective occupied considerable attention, from the artistic point of view by the immortal Leonardo da Vinci (1500), and from the theoretical point of view by geometers such as Desargues. The eighteenth century is well termed the golden age of theoretical perspective, for just at that period when the methods of orthogonal projection exhibited their most robust growth, central projection attained a stage of full development, and could claim as its own some of the great names of the age. First among them comes s'Gravesande (1688–1742), to whom the representation of a line by its trace and vanishing point is due. His *Essai de Perspective* builds up on the theory of central projection the framework of a complete and connected system of descriptive geometry. No less worthy a place is that held by the great English mathematician, Brook Taylor (1685–1731), known to the world by his works on analysis. His *New Principles of Linear Perspective* is the foundation on which all modern works on central projection are based, and in it appears, for the first time, the now universally accepted terminology. Lastly, there is the name of Lambert (1728–1777), whose *Freye Perspective* represents the fullest development of perspective geometry in this age most favourable to its growth. The nineteenth century was far less productive than the eighteenth, though in the history of perspective the name of Fiedler cannot be passed over without mention.

More recent developments have been connected with the practical applications of the methods of orthogonal projection and of perspective, and with the closely allied art of photogrammetry rather than with the theoretical aspects of these systems of descriptive geometry.

CHAPTER II

THE STRAIGHT LINE AND PLANE IN ORTHOGONAL PROJECTION

9. In the following sections, Monge's Method is applied to the solution of several important problems, which are not only of intrinsic interest as illustrating the methods of this branch of descriptive geometry, but are of continual application in more advanced work when space curves and curved surfaces are dealt with. Much of what is to follow depends on the following *lemmas*:—

α. If two lines in space intersect, the straight line joining the point of intersection of their horizontal projections to the point of intersection of their vertical projections is perpendicular to the ground line, and conversely.

β. If two planes are parallel, their traces are respectively parallel. The converse is true *provided the traces are not parallel to the ground line.*

γ. If two lines in space are parallel, their vertical projections and likewise their horizontal projections are respectively parallel, and conversely.

δ. If two lines are perpendicular to one another, their projections, on a plane parallel to either of them, are perpendicular, and conversely.

ϵ. If two given straight lines are perpendicular to one another, a plane can be drawn through one of them perpendicular to the other. The traces of this plane pass through the traces of the first given line and are perpendicular to the projections of the second given line. The condition for the existence of such a plane is that these traces should intersect on the ground line, which is therefore the condition that the two given straight lines should be perpendicular to one another.

ξ. If a line is perpendicular to a plane, the projections of the line are respectively perpendicular to the traces of the plane, and conversely.

[Consider the vertical plane π through the line. It is perpendicular to the horizontal plane of projection and also to the given plane, and, in consequence, is at right angles to the horizontal trace of the given plane. But the plane π contains the horizontal projection of the given line, and hence the

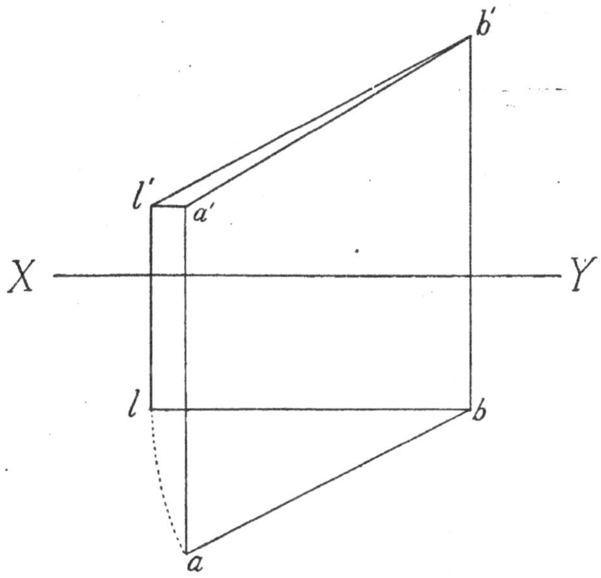

FIG. 9.

horizontal projection of the line is perpendicular to the horizontal trace of the plane.]

10. Problem I.—*To find by construction the length of a line whose horizontal and vertical projections are given.*

Let ab and $a'b'$ be the given horizontal and vertical projections respectively. Imagine the line itself to be rotated in space about the vertical bB as axis (fig. 1) until it is parallel to the vertical plane of projection. Let its projections in the new position be bl and $b'l'$ respectively (fig. 9). Then since the line, in its new position, is parallel to the vertical plane of projection, bl will be parallel to the ground line, and in length bl will be equal to ba. Hence the position of l is known and can be constructed.

Now since $b'l'$ is the vertical projection of the line in its new position, l' is at the same height above the ground line as a', and since l and l' are the projections of the same point, ll' is at right angles to the ground line. Hence the point l' is also known and can be constructed.

But since the actual line in space is now parallel to the vertical plane of projection, its real length is equal to the length of its vertical projection.

Hence $b'l'$ gives the required length of the line.

Examples: *

I. $a=(-12{\cdot}4,\ -15{\cdot}7)$, $a'=(-12{\cdot}4,\ 21{\cdot}3)$, $b=(17{\cdot}9,\ -5{\cdot}4)$, $b'=(17{\cdot}9,\ 12{\cdot}2)$.

II. $a=(-9{\cdot}8,\ -3{\cdot}4)$, $a'=(-9{\cdot}8,\ 16{\cdot}5)$, $b=(16{\cdot}3,\ -19{\cdot}4)$, $b'=(16{\cdot}3,\ 8{\cdot}6)$.

III. $a=(-17{\cdot}2,\ -9{\cdot}1)$, $a'=(-17{\cdot}2,\ 8{\cdot}7)$, $b=(20{\cdot}7,\ 2{\cdot}3)$, $b'=(20{\cdot}7,\ 23{\cdot}8)$.

11. Problem II.—*Given a point (p, p') and a straight line $(ab, a'b')$, to draw the line through (p, p') parallel to the given line.*

Since the given line and the required line are parallel, their horizontal and vertical projections must be parallel. Moreover, since the required line passes through the point (p, p'), its horizontal projection must pass through p and its vertical projection through p'. The required construction thus becomes obvious.

Example: $a=(-18{\cdot}1,\ -20{\cdot}6)$, $a'=(-18{\cdot}1,\ 3{\cdot}8)$, $b=(14{\cdot}7,\ -18{\cdot}4)$, $b'=(14{\cdot}7,\ 15{\cdot}8)$, $p=(5{\cdot}4,\ -12{\cdot}3)$, $p'=(5{\cdot}4,\ 3{\cdot}7)$.

12. Problem III.—*Given the projections $(ab, a'b')$ of a line, to find its traces, and conversely.*

Consider the vertical trace c' (fig. 10). It must lie in the

* The examples given are adapted for work on an imperial sized drawing board. The specified points are to be set off with respect to the ground line as x-axis and a vertical line near the centre of the paper as y-axis. All lengths are given in cm.

vertical projection $a'b'$ (produced if necessary). Further, the horizontal projection of c', being the horizontal projection of the point of intersection of the given line and the vertical plane of projection, must be the intersection of their horizontal projections. Hence the horizontal projection of c' is the intersection of ab with the ground line.

The vertical trace c' is therefore the intersection of $a'b'$ (produced if necessary) with a line cc' drawn at right angles to the ground line through c, the intersection of ab with the ground line. Similarly, if d' is the point in which $a'b'$ meets the ground

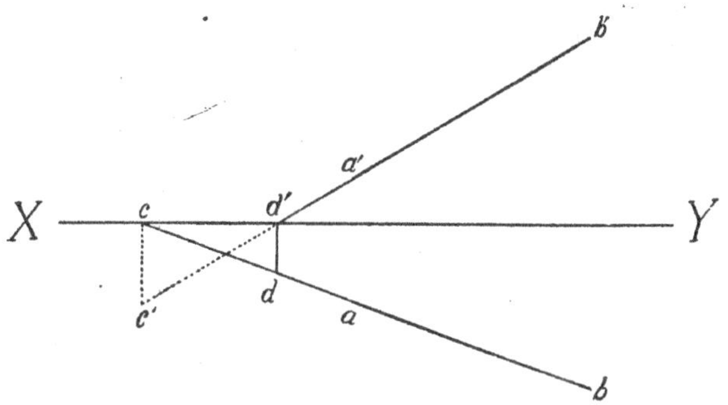

FIG. 10.

line, and if $d'd$ is drawn at right angles to the ground line to meet ab in d, then d is the horizontal trace of the given line.

The solution of the converse problem, *to find the projections of a line whose traces d and c′ are given* is now obvious, for it simply amounts to dropping the perpendiculars dd' and $c'c$ on to the base line. The required projections are then cd and $c'd'$.

It is well to note the importance of the *fundamental trapezium, cc′d′d,* which reappears in almost every problem connected with the straight line and plane.

13. Problem IV.—*Given a plane whose traces are aB and Bc′, and a point whose projections are p and p′, to find the traces of a plane through (p, p′) parallel to the given plane.*

Since the required plane is to be parallel to the given plane, its traces will be parallel to aB and Bc' respectively. Consider now a horizontal line drawn through (p, p') in the required plane.

Its projections are a line through p' parallel to the ground line (vertical projection), and a line through p parallel to aB (horizontal projection). Now obtain, by the method indicated in Problem III., the vertical trace l' of this line; l' is a point in the vertical trace of the required plane, and therefore the vertical

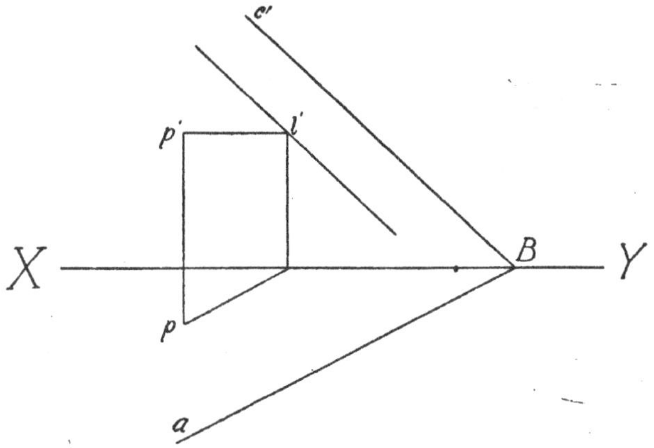

FIG. 11.

trace of the required plane is a line drawn through l' parallel to $c'B$. Since the horizontal and vertical traces meet on the ground line, the required horizontal trace can now also be constructed.

Example: $a=(16{\cdot}3,\ -19{\cdot}6)$, $B=(-27{\cdot}8,\ 0)$, $c'=(14{\cdot}6,\ 22{\cdot}9)$, $p=(15{\cdot}1,\ -7{\cdot}6)$, $p'=(15{\cdot}1,\ 19{\cdot}4)$.

14. Problem V.—*To draw the traces of the plane which passes through three given points* (a, a'), (b, b'), *and* (c, c').

Join ab, ac, and bc, and also $a'b'$, $a'c'$, and $b'c'$. The lines of which these are the projections lie wholly in the required plane, and hence their vertical traces lie in the vertical trace of the required plane. If, therefore, e' is the vertical trace of $(ab, a'b')$, f' of $(bc, b'c')$, and g' of $(ca, c'a')$, then e', f', and g' lie on a line which is the required vertical trace. The horizontal trace may be found in like manner.

The problem, *to draw the traces of the plane which passes through two intersecting straight lines whose projections are given*, can obviously be reduced to the above, for it amounts simply to drawing the traces of a plane which passes through the point of intersection of the lines and through one other chosen point on each of them. The same applies to the problem, *to draw the traces of the plane which passes through a point and a straight line whose projections are given.*

Example: $a=(-21{\cdot}6,\ -4{\cdot}7)$, $a'=(-21{\cdot}6,\ 19{\cdot}5)$, $b=(25{\cdot}8,\ -8{\cdot}4)$, $b'=(25{\cdot}8,\ 19{\cdot}5)$, $c=(18{\cdot}3,\ -4{\cdot}7)$, $c'=(18{\cdot}3,\ 11{\cdot}8)$.

15. Problem VI.—*Given the horizontal projection of a line, and the traces of a plane in which it lies, to find its vertical projection.*

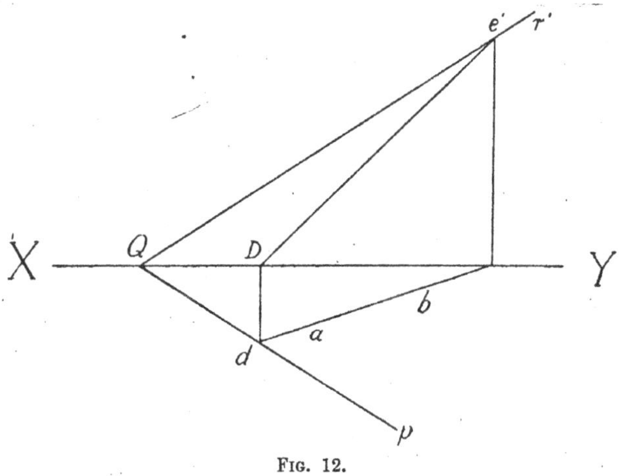

FIG. 12.

Let ab be the horizontal projection of the given line, and pQ and Qr' the traces of the given plane. Then the vertical trace of the line must lie on Qr', and must have as its horizontal projection the point in which ab meets the ground line. Hence the vertical trace e' can be constructed. The horizontal trace of the line can also be found, for it is simply the point of intersection d of pQ and ab produced. If, therefore, D be the foot of the perpendicular drawn from d to the ground line, the required vertical projection will lie along De'.

This construction, as well as others of similar nature, may sometimes appear to fail owing to peculiar features of the data, but such cases may nearly always be regarded as limiting cases of a perfectly soluble form of the problem. Thus, for example, if the horizontal projection ab be parallel to the horizontal trace Qp, it will be impossible to find their point of intersection d, and the corresponding point D. In this case, however, D may be regarded as a point situated at an infinitely great distance along the ground line, so that the required vertical trace $e'D$ is now a line through e' parallel to the ground line.

Example: $a=(-3{\cdot}6, 2{\cdot}9)$, $b=(17{\cdot}8, -12{\cdot}3)$, $p=(15{\cdot}0, -13{\cdot}8)$, $Q=(-8{\cdot}6, 0)$, $r'=(19{\cdot}0, 12{\cdot}6)$.

16. Problem VII.—*To find the line of intersection of two planes whose traces are given.*

It is evident that the horizontal trace of the required line will be the intersection of the horizontal traces of the two planes, and that its vertical trace will be the intersection of their vertical traces. The traces of the line of intersection being thus known, its projections may also be found.

Such limiting cases as arise may be dealt with in the manner indicated in the preceding section, with the exception of the notable exceptional case in which all the traces are parallel to the ground line. In this case it is necessary to consider the projections of the given planes on an auxiliary plane inclined at an angle to the two planes of projection. The manner in which this is carried out is indicated below.

17. Introduction of a Third Plane of Projection.—As a rule, the auxiliary plane may be taken to be perpendicular to the ground line and hence plays the part of a new vertical plane of projection, on which the traces of the two planes are to be found. This plane is then rabatted back on to the original vertical plane, and then together with the latter is rabatted down on to the horizontal plane, thus giving rise to a plane diagram (fig. 13). This plane diagram is divided into four quadrants by the ground lines XOY and YOZ, the region XOY containing the horizontal projection and the regions XOZ and YOZ the old and new vertical projections respectively. The region YOY serves to connect up the new vertical projection with the horizontal projection, for since the two lines OY originally coincided, they really represent one and the same straight line and are therefore congruent with each other. This is denoted by connecting up corresponding points on these two lines in the plane diagram with quadrants of circles whose centre lies at *O*.

Now let *ab* and *a′b′* be the horizontal and vertical traces of the first plane as originally represented, and *cd* and *c′d′* those of the second plane, and let these traces (produced if necessary) cut *YOZ* in *P*, *Q*, *R*, and *S* respectively. Transfer the points *P* and *R* to their corresponding positions on *XOY* and join *PQ* and *RS*, which are the traces of the given planes on the new vertical

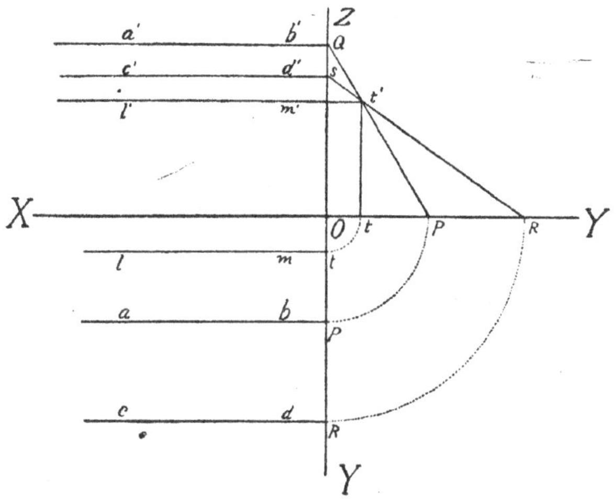

FIG. 13.

plane. Let their intersection be *t′*. Then the vertical projection of the required line of intersection is the line *l′m′* drawn through *t′* parallel to the ground line. The horizontal projection *t* of *t′* is a point in the horizontal projection *lm* of the required line, which can now be constructed.

> *Example:* *ab* and *cd* are respectively 4·5 and 18·6 cm. below and *a′b′* and *c′d′* respectively 10·8 and 14·7 cm. above the ground line.

18. Problem VIII.—*To construct the point of intersection of a given line and a given plane.*

Take any auxiliary plane through the given line and find its intersection with the given plane. Where the line of intersection meets the given line is the required point.

Let (ab, $a'b'$) be the given line and pQr' the given plane. The simplest auxiliary plane to take is the vertical plane through ab; its horizontal trace will be ab and its vertical trace cc', where c is the intersection of ab produced with the ground line, and cc' is drawn at right angles to the ground line at c. The traces of the intersection of this auxiliary plane with the given plane are d and c', and its projections are cd and $c'd'$. Hence the vertical

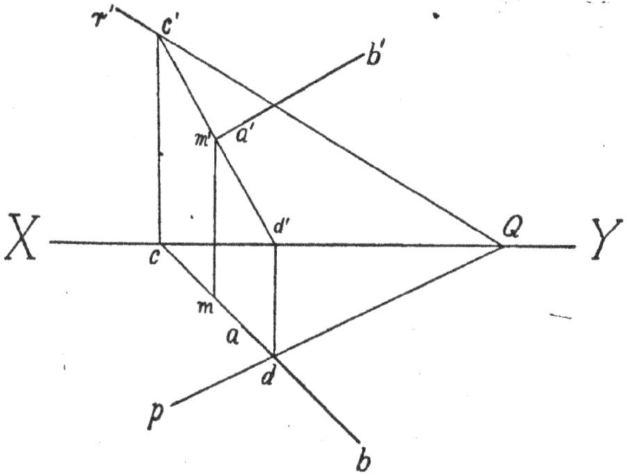

FIG. 14.

projection m' of the required point is the intersection of $c'd'$ with $a'b'$, and if $m'm$ is drawn perpendicular to the ground line to meet ab in m, then m is the corresponding horizontal projection.

Example: $a = (10{\cdot}5, \ -8{\cdot}5)$, $a' = (10{\cdot}5, \ 11{\cdot}5)$, $b = (2{\cdot}5, -2{\cdot}5)$, $b' = (2{\cdot}5, 6{\cdot}7)$, $p = (4, -7{\cdot}6)$, $Q = (20{\cdot}5, 0)$, $r' = (-2{\cdot}7, 17{\cdot}5)$.

19. Problem IX.—*To find the plane which passes through a given point and a given straight line.*

The general method of solution is as follows:—Join the given point to some convenient point of the line so as to obtain a second line situated in the required plane. The vertical traces of these two lines lie on the vertical trace of the required plane, which is thus determined. The corresponding horizontal trace is obtained in like manner.

Let *ab* and *a'b'* be the horizontal and vertical projections of the given line, *b* and *a'* being the corresponding traces, and let (*p*, *p'*) be the given point. The most convenient point in the line (*ab*, *a'b'*) to which (*p*, *p'*) may be joined is the point at infinity, in which case the new line is parallel to the given line Through *p* and *p'*, therefore, draw lines *pl* and *p'k'* parallel to *ab* and *a'b'* respectively, and find the traces *l* and *k'* of the line

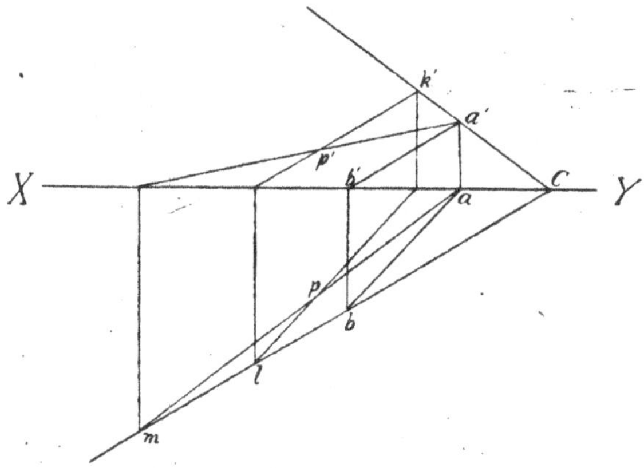

FIG. 15.

they represent. Then *lbC* and *k'a'C* are the traces of the required plane.

Instead of the point at infinity the point (*a*, *a'*), that is to say, the vertical trace of the given line, may be taken as the point to which to join (*p*, *p'*). The horizontal trace *m* of the line (*ap*, *a'p'*) is a point on the horizontal trace *mbC* of the required plane, and when this is drawn, the corresponding vertical trace *Ca'* can also be constructed. The horizontal trace (*b*, *b'*) would serve equally well as the chosen point, giving ultimately a second point on the vertical trace of the required plane.

> *Example:* $p=(8, -9)$ $p'=(8, 4{\cdot}5)$. Vertical trace of given line $(13{\cdot}5, 15)$, horizontal trace $(28, -22)$.

20. Problem X.—*Through a given point to draw a line to meet two given straight lines.*

Draw a plane through the given point and the first line, and a second plane through the given point and the second line. Then the intersection of these planes is evidently the required line.

Otherwise, draw a plane through the given point and one of the straight lines, and determine the point where this plane cuts the other straight line. The line joining this point to the given point is the line required.

> *Example:* Vertical trace of first line (10·3, 8·6), horizontal trace (−3·6, −27); vertical trace of second line (−10·8, 10·4), horizontal trace (14·7, −11); vertical projection of given point (2·5, 6·1), horizontal projection (2·5, −10·7).

21. Problem XI.—*To find the shortest distance from a given point* (a, a') *to a given plane* (pQr').

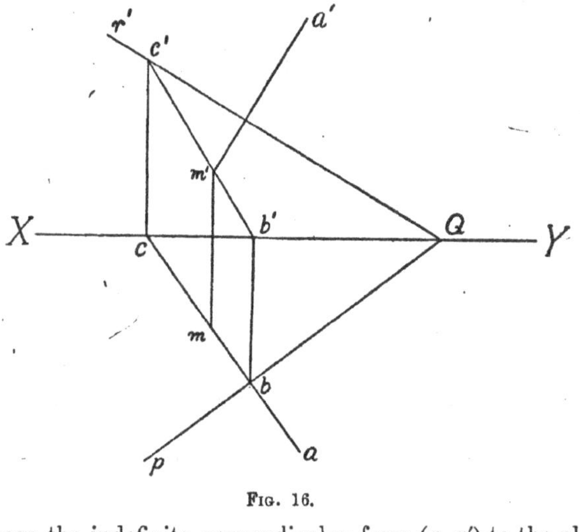

FIG. 16.

Draw the indefinite perpendicular from (a, a') to the plane; its projections will be the perpendiculars am, $a'm'$ from a and a' respectively on to the corresponding traces of the plane. The distance of the foot of this perpendicular from (a, a') is evidently the distance required. The foot of this perpendicular is found by drawing the vertical plane acc' through the given line and obtaining the intersection of this plane with the given plane. The point where this line meets the given line will be the required point. Thus if b and c' are the traces of the line of intersection of the given plane with the vertical plane through the perpendicular line, ab and $c'b'$ are the horizontal and vertical projections respectively of this line of intersection, b' being on the ground line. Consequently, if the perpendicular through a' meets $c'b'$ in m', m' is the vertical projection of the foot of the perpendicular, and the horizontal trace m can now readily be obtained. Hence am and $a'm'$ are respectively the horizontal and vertical projections of the required perpendicular, whose true length can now be found.

Example: $a=(-18{\cdot}2,\ -17{\cdot}7)$, $a'=(-18{\cdot}2,\ 19{\cdot}6)$, $p=(23{\cdot}4,\ -20{\cdot}0)$, $Q=(-27{\cdot}8,\ 0)$, $r'=(18{\cdot}5,\ 16{\cdot}7)$.

22. Problem XII.—*To find the shortest distance from a given point* (c, c') *to a given straight line* $(ab, a'b')$.

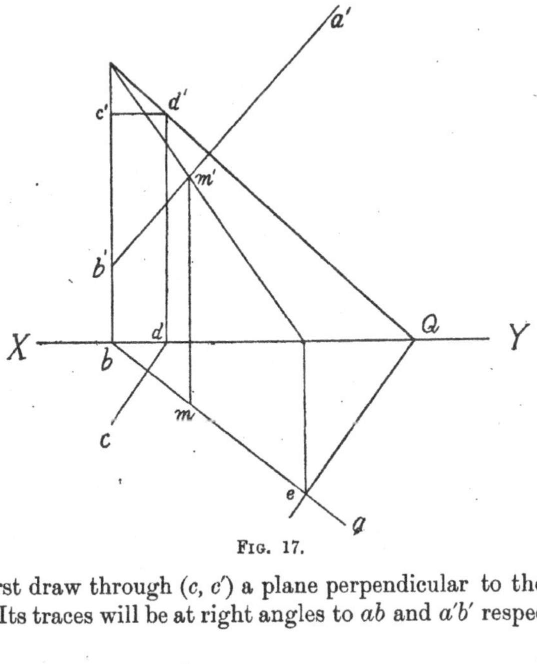

FIG. 17.

First draw through (c, c') a plane perpendicular to the given line. Its traces will be at right angles to ab and $a'b'$ respectively, and hence they can be found if a point through which one of them passes is known. In order to find such a point, consider a horizontal line through (c, c') lying in the required plane. The horizontal projection of this line will be parallel to the horizontal trace of the plane, that is to say, at right angles to the horizontal projection of the given line. The vertical projection of this line will be horizontal, and d', its vertical trace, will lie on the vertical trace of the required plane. Thus $d'Q$, drawn at right angles to $a'b'$, is the vertical trace of the required plane, and Qe, drawn perpendicular to ab, is the corresponding horizontal trace.

The problem is now reduced to that of finding the intersection (m, m') of a given plane eQd' with a given line $(ab', a'b')$. This is done by drawing the vertical plane through $(ab, a'b')$; its traces are ab and bb' where bb' is perpendicular to the ground line.

Example: $a=(23{\cdot}7,\ -17{\cdot}4)$, $a'=(23{\cdot}7,\ 26{\cdot}5)$, $b=(-12{\cdot}9,\ -4{\cdot}3)$, $b'=(-12{\cdot}9,\ 8{\cdot}6)$, $c=(-6{\cdot}5,\ -9{\cdot}7)$, $c'=(-6{\cdot}5,\ 15{\cdot}8)$.

23. Problem XIII.—*To find the shortest distance from a given point (c, c') to a given straight line $(ab, a'b')$, by a method of rabatting, a and b' being the traces of the given line.*

This is an alternative method of solving the last problem. Draw a plane through the point (c, c') and the line $(ab, a'b')$. This is done by joining (c, c') to the point (a, a') and then finding the vertical trace d' of this new line. Then b' and d' are two points in the vertical trace $b'd'Q$ of the required plane, which can now be drawn. The corresponding horizontal trace is Qa, which can also be drawn. Now rabat this plane about its horizontal trace Qa. During rabatment the point (b, b') will remain in a vertical plane whose horizontal trace is the line bf drawn through b at right angles to Qa. The distance of this point from Q will remain constant, and consequently, if a circle of centre Q and radius Qb' be drawn to meet bf in b'', then b'' will be the point into which (b, b') is rabatted. The given straight line $(ab, a'b')$ will therefore be rabatted into the position ab''. Now if d'' is the rabatment of d', then d'' will lie on Qb'' at a distance from Q equal to Qd'. Thus d'' can be found, and hence ad'', the rabatment of $(ac, a'c')$, can be constructed. To obtain the point into which (c, c') is rabatted, draw cc'' perpendicular to Qa, meeting ad'' in c'', which is the required rabatment of (c, c'). Hence to obtain, in its rabatted position, the perpendicular from (c, c') on $(ab, a'b')$, draw the perpendicular from c'' on ab''. The length of this perpendicular $(c''m'')$ is the actual length required. If the projections of this perpendicular are required, it is necessary to rabat the point m'' back again into its position in space. For this purpose draw a perpendicular from m'' on to the line Qa. Then the point m, in which this perpendicular meets ab, will be the horizontal projection of the foot of the perpendicular from (c, c')

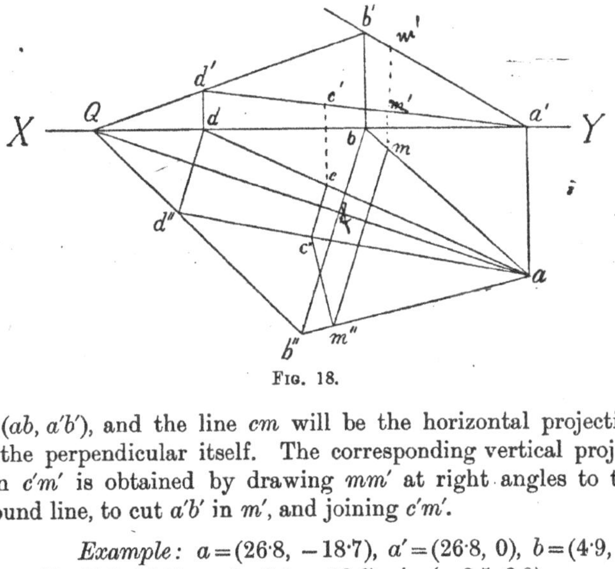

FIG. 18.

on $(ab, a'b')$, and the line cm will be the horizontal projection of the perpendicular itself. The corresponding vertical projection $c'm'$ is obtained by drawing mm' at right angles to the ground line, to cut $a'b'$ in m', and joining $c'm'$.

Example: $a=(26{\cdot}8, -18{\cdot}7)$, $a'=(26{\cdot}8, 0)$, $b=(4{\cdot}9, 0)$, $b'=(4{\cdot}9, 15{\cdot}3)$, $c=(-3{\cdot}5, -12{\cdot}6)$, $c'=(-3{\cdot}5, 2{\cdot}8)$.

24. This method of rabatting is of considerable value as a means of solving problems in which it is required to find the true magnitude of some element of the figure in space. Thus suppose it is required to find the radius of the circle which passes through three given points in space. This problem is immediately solved by finding the plane which passes through these three points, and rabatting this plane on to the horizontal plane of projection. The three points now appear in their true relative positions, and the centre and radius of the required circle can be found by the methods of plane geometry. In the following sections the methods of finding the magnitudes of angles in space, by means of rabatting, will be illustrated.

25. Problem XIV.—*To find the angle between two given straight lines.*

If the two straight lines do not intersect, then they may be replaced by straight lines OA and OB drawn through any convenient point O parallel to the given lines, and the angle

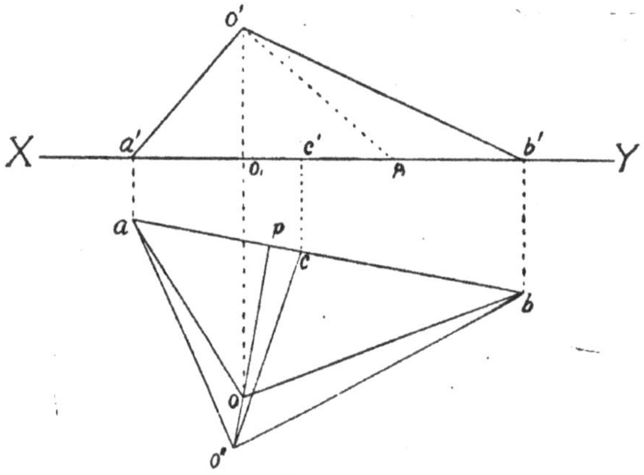

FIG. 19.

AOB considered as the angle between them. The case of two intersecting straight lines will therefore alone be considered.

Let $(ao, a'o')$ and $(bo, b'o')$ be the two given straight lines, intersecting in (o, o'), and let a and b be their respective horizontal traces, so that a' and b' lie on the ground line. Then ab is the horizontal trace of the plane containing the two given lines, and, consequently, if this plane be rabatted about ab on to the horizontal plane, and if o'' is the rabatment of (o, o'), then the angle between ao'' and $o''b$ will be the angle required. During the rabatment, the horizontal trace o moves along the line po perpendicular to ab, and since o'' is the rabatted position of (o, o'), the length po'' will be the true altitude of the triangle whose vertex is (o, o') and base $(ab, a'b')$. But, obviously, the vertical projection of the altitude is $o'o_1$, where o_1 is the intersection of oo' with the ground line, and the corresponding horizontal projection is op. If, therefore, a length o_1p_1 equal to op be measured off along the ground line, then $o'p_1$ is the true altitude of the triangle, and if a length po'' equal to $o'p_1$ be measured along po produced, then o'' is the required rabatment of (o, o'). The angle $ao''b$ is now the angle required, and its magnitude can be found by the usual methods.

Example: $a = (-13{\cdot}0,\ -5{\cdot}8)$, $a' = (-13{\cdot}0,\ 0)$, $b = (10{\cdot}8,\ -10{\cdot}3)$, $b' = (10{\cdot}8,\ 0)$, $o = (-9{\cdot}8,\ -12{\cdot}3)$, $o' = (-9{\cdot}8,\ 9{\cdot}7)$.

26. Several new problems arise as corollaries of the last and can now easily be treated. Thus suppose it is required *to draw the bisector of the angle formed by the lines* $(ao, a'o')$ *and* $(bo, b'o')$. Let $ao''b$ be the angle in its rabatted position, bisect it by the line $o''c$ meeting ab in c, and then rabat it back into its original position in space. During this latter process the point c, the horizontal trace of the bisecting line, remains fixed, together with the corresponding vertical trace c' which lies on the ground line. Hence oc and $o'c'$, the projections of the required bisector, can be drawn. Again, suppose it is required *to construct the angle between a given straight line and a given plane.* The angle considered is the angle between the given line and its projection on the given plane, so that the problem amounts to constructing this projection and finding the angle between it and the given line. Take any point on the given line and find the perpendicular drawn from it on to the given plane. The line joining the foot of this perpendicular to the point where the given line meets the plane is the required line, and hence the problem can now be solved. The problem of the next section is yet another example of the same method.

Example: $(ab, a'b')$ is the given line and pQr' the given plane where $a = (-14{\cdot}6,\ -15{\cdot}8)$, $a' = (-14{\cdot}6,\ 7{\cdot}9)$, $b = (3{\cdot}5,\ -7{\cdot}2)$, $b' = (3{\cdot}5,\ 7{\cdot}2)$, $p = (-18{\cdot}5,\ -16{\cdot}7)$, $Q = (15{\cdot}3,\ 0)$, $r' = (-21{\cdot}9,\ 19{\cdot}8)$.

27. Problem XV.—*To find the angle between two planes whose traces are given.*

If a plane be drawn through any convenient point of the line of intersection of the two given planes, and at right angles to these two planes, then the angle between the lines in which the new plane cuts the two given planes is the angle required.

Let *pRq′* and *pSq′* be the required planes, and let *pq*, the horizontal projection of their line of intersection, be drawn. Then if *c* is the horizontal projection of the chosen point on the line of intersection, the horizontal trace of the auxiliary plane will be the line *ab* drawn through *c* at right angles to *qp*. Hence if *a* and *b* be the points where this line cuts the horizontal traces of the given planes, the problem resolves itself

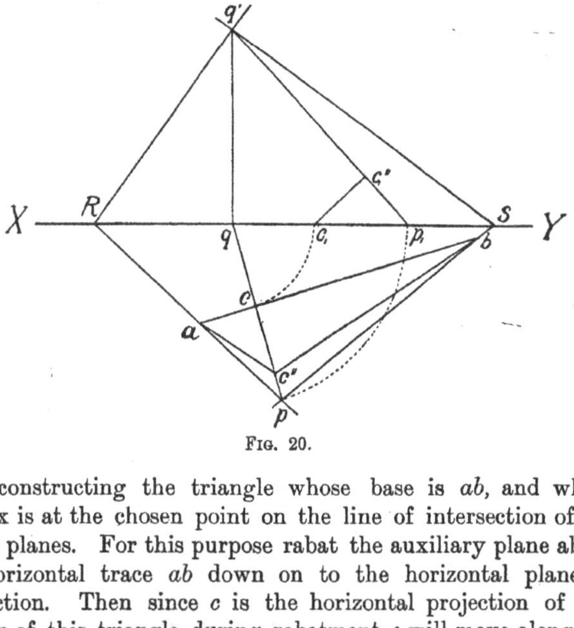

FIG. 20.

into constructing the triangle whose base is *ab*, and whose vertex is at the chosen point on the line of intersection of the given planes. For this purpose rabat the auxiliary plane about its horizontal trace *ab* down on to the horizontal plane of projection. Then since *c* is the horizontal projection of the vertex of this triangle, during rabatment *c* will move along the line *qp* at right angles to *ab*, and the altitude of the triangle will ultimately lie along *qp*. To find the true length of this altitude, rabat the vertical plane through *qp* about its vertical trace back on to the vertical plane of projection, so that the required altitude, being originally vertical, will appear in its true length. The points *c* and *p* will be rabatted on to the ground line so that if c_1 and p_1 are their new positions, qc_1 and

qp_1 are respectively equal to qc and qp. Now, since the required altitude was originally perpendicular to qp, the rabatted altitude is obtained by drawing c_1c_1'' at right angles to $q'p_1$, which gives the true length required. Then, if a length cc'' equal to c_1c_1'' is measured off along qp, the angle $ac''b$ will be the angle required.

Example: $p=(4\cdot1,\ -21\cdot5)$, $R=(-27\cdot5, 0)$, $q'=(12\cdot9, 18\cdot6)$, $S=(8\cdot3, 0)$.

28. Problem XVI.—*Given the three faces of a solid angle (angles α, β, γ), to find the angles between the planes.*

Among all the practical applications of descriptive geometry,

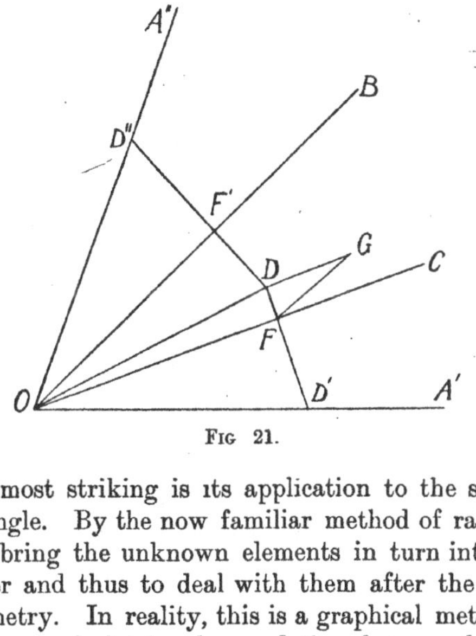

FIG 21.

one of the most striking is its application to the solution of a trihedral angle. By the now familiar method of rabatting it is possible to bring the unknown elements in turn into the plane of the paper and thus to deal with them after the methods of plane geometry. In reality, this is a graphical method for the solution of spherical triangles, and the above problem merely amounts to finding the three angles of a spherical triangle whose three sides are given.

Let OA, OB, and OC be the space positions of the three edges of the given trihedral angle so that $\angle BOC=\alpha$, $\angle COA=\beta$, $\angle AOB=\gamma$.

Now rabat the two faces COA and BOA on to the plane of the face BOC, their new positions being COA' and BOA'', so that OA' and OA'' represent the single line OA in space, and $\angle COA' = \beta$, and $\angle BOA'' = \gamma$.

Consider now some point D' on OA' and let D'' be the corresponding point on OA'' so that $OD' = OD''$. Draw $D'F$ at right angles to OC and $D''F'$ at right angles to OB, meeting $D'F$ produced in D. Now if the faces COA' and BOA'' be rebatted back again to their former positions in space, D' will, during the process of erection, remain in the vertical plane through $D'F$ and D'' in the vertical plane through $D''F'$, so that D will be the horizontal projection of the actual position of the point on the third edge corresponding to D' and D''. Thus it follows that in the vertical face of which DF is a horizontal projection there will be a right-angled triangle of which the base is DF and the hypotenuse FD'. Rabat this triangle about DF so that it becomes DGF where DG is perpendicular to DF and $FG = FD'$. Then the angle DFG is really the angle between two lines in the planes of the angles α and β which are both perpendicular to the edge OF, and is therefore the required angle C.

This construction admits of a simple analytical verification, as follows. If α, β, γ are the sides of a spherical triangle, and A, B, C the corresponding angles, then

$$\cos\gamma = \cos\alpha\cos\beta + \sin\alpha\sin\beta\cos C.$$

If now $\qquad OD = \rho,\ OD' = r, \text{ and } \angle DOF = \theta,$

then evidently $\qquad \rho\cos\theta = OF = r\cos\beta,$

and $\qquad \rho\cos(\alpha - \theta) = OF' = r\cos\gamma.$

Hence
$$\rho\sin\alpha\sin\theta = r\cos\gamma - r\cos\beta\cos\alpha$$
$$= r\sin\alpha\sin\beta\cos C,$$

i.e. $\qquad \rho\sin\theta = r\sin\beta\cos C$

or $\qquad DF = FD'\cos C = FG\cos C.$

The angle DFG is therefore the required angle C.

The angle B is obtained from the diagram in the same way. In order to find the angle A, consider a plane drawn through a point on the third edge whose projection is D, perpendicular to the third edge.

Since OA' is a rabatment of the third edge, the intersection of this new plane with the face β will be a perpendicular to OA' at D'. If this perpendicular be $D'N$, the new plane will meet OC in N, and if $D''M$ be drawn perpendicular to OA'' the new plane will meet OB in M, and thus MN will be the actual intersection of the new plane with the plane *BOC*. A triangle therefore exists in the new plane whose sides when rabatted on to the plane *BOC* are $D''M$, MN, and ND'. Rabat this triangle around MN, so that it becomes MPN where $MP = MD''$ and

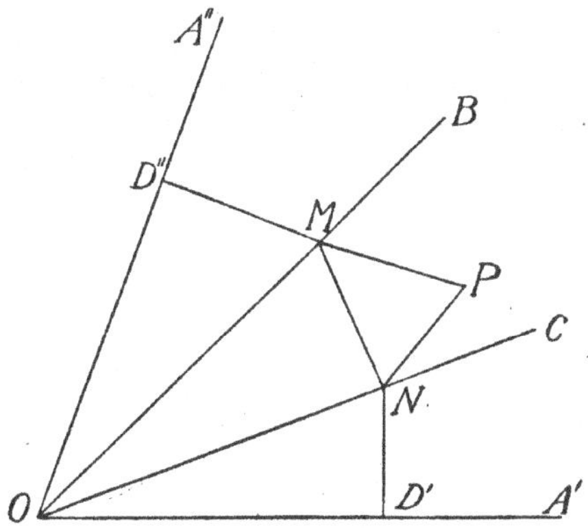

FIG. 22.

$NP = ND'$. Now PM and PN are the rabatments of two lines perpendicular to the third edge, drawn through a point on it, and hence the angle MPN is the required angle A.

Examples: I. $\alpha = 60°$, $\beta = 60°$, $\gamma = 45°$.
II. $\alpha = 22\frac{1}{2}°$, $\beta = 90°$, $\gamma = 30°$.
III. $\alpha = 120°$, $\beta = 45°$, $\gamma = 75°$.

29. Problem XVII.—*Given two faces and one angle of a trihedral angle, to find the other face and angles.*

Suppose α, β, and C, that is to say, two faces and the included angle are given. Draw the angle BOC equal to α and the angle COA' equal to β (fig. 21), and through D' some convenient point on OA', draw $D'F$ perpendicular to OC. At F draw FG', making an angle C with $D'F$ produced. Now take FG to be equal to FD' and from G draw GD perpendicular to $D'F$ produced, so that D is as before the projection of a point on the third edge. Now draw DF' perpendicular to OB and on it take D'' such that OD'' is equal to OD'. Then $D''OF'$ is the required third side γ, and the remaining elements of the trihedral angle can be found as before.

The case in which the given angle is opposite one of the given faces can be treated in a similar manner.

Examples: I. $\alpha=60°$, $\beta=45°$, $C=75°$.
II. $\alpha=30°$, $\beta=90°$, $A=60°$.

30. Problem XVIII.—*To reduce a given angle to the horizon.*

Suppose two lines are given inclined at angles β and γ to the vertical and at an angle α to each other, it is required to find the angle α' which their horizontal projections make with

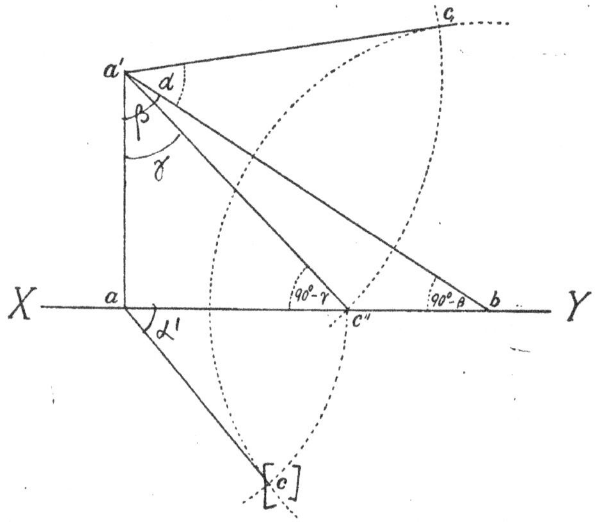

FIG. 23.

each other. This problem is of considerable importance in surveying and is known as "reducing the given angle to the horizon." It is evident that α, β, and γ may be taken as the three faces of a trihedral angle, so that the problem consists in finding the angle enclosed by the faces β and γ, and is therefore soluble by the method given in Problem XVI. It may also be solved directly in the following manner:—

Let $aa'b$ be the given angle β, the line ab being horizontal. Take ab to be the ground line and the plane of $aa'b$ to be the vertical plane of projection. Then if c be the point in which the other given line meets the horizontal plane of projection, the angle bac will be the required angle α'. Now if $a'c''$ be the rabatment of this other line on the plane $aa'b$, the angle $aa'c''$ will be γ, and hence c'' is known. If now this line be rabatted back again to its original position in space, the point c in which it meets the horizontal plane describes a circle of radius ac'' about a as centre. Consider the distance of this point c from b. The line bc is the base of a triangle whose sides are equal to $a'b$ and $a'c''$ and whose vertical angle is a. At a', therefore, draw a line $a'c_1$ making an angle a with $a'b$ and cut off a length $a'c_1$ equal to $a'c''$. With centre b and radius bc_1 describe a circle cutting the circle of centre a and radius ac'' in c, which is the horizontal trace of the second given line. The angle cac'' is now the required angle α'.

Example: $\beta = 67\frac{1}{2}°$, $\gamma = 56°$, $a = 37\frac{1}{2}°$.

EXERCISES ON CHAPTER II

1. On a given straight line (ab, $a'b'$), to find the point (c, c') which is at a given distance d from the extremity (a, a').

Example: $a = (-17\cdot1, -14\cdot8)$, $a' = (-17\cdot1, 12\cdot5)$, $b = (16\cdot0, -0\cdot6)$, $b' = (16\cdot0, 21\cdot3)$, $d = 17\cdot6$.

2. Given two parallel straight lines (ab, $a'b'$) and (cd, $c'd'$), to find the distance between them.

Example: $a = (-11\cdot4, -12\cdot7)$, $a' = (-11\cdot4, 16\cdot8)$, $b = (15\cdot3, -2\cdot8)$, $b' = (15\cdot3, 2\cdot4)$, $c = (-9\cdot6, -5\cdot6)$, $c' = (-9\cdot6, 21\cdot9)$.

3. Through a given straight line (ab, $a'b'$) to draw a plane parallel to a given straight line (cd, $c'd'$), and hence to find the shortest distance between the two given straight lines.

Example: $a = (-14\cdot9, -2\cdot6)$, $b = d = (18\cdot4, -7\cdot8)$, $c = (-14\cdot9, -12\cdot5)$, $a' = c' = (-14\cdot9, 7\cdot4)$, $b' = (18\cdot4, 20\cdot7)$, $d' = (18\cdot4, 1\cdot9)$.

4. A straight line whose vertical projection is $a'b'$ passes through a given point (a, a') and makes a given angle α with the horizontal plane, to construct its horizontal projection.

Example: $a = (-10\cdot0, -6\cdot5)$, $a' = (-10\cdot0, 9\cdot0)$, $b' = (15\cdot0, 19\cdot5)$, $\alpha = 15°$.

5. Through a given point (p, p') to draw a straight line intersecting a given straight line (ab, $a'b'$) and making an angle α with it.

Example: $a = (-9\cdot8, -3\cdot2)$, $a' = (-9\cdot8, 0)$, $b = (-2\cdot1, 0)$, $b' = (-2\cdot1, 11\cdot5)$, $p = (18\cdot7, -2\cdot6)$, $p' = (18\cdot7, 16\cdot2)$, $\alpha = 30°$.

6. Given the horizontal projection p of a point (p, p') in a given plane qRs', to draw a line through (p, p') making an angle α with the line of steepest slope in the plane.

Example: $p=(10{\cdot}5, -6{\cdot}8)$, $q=(11{\cdot}7, -8{\cdot}4)$, $R=(-14{\cdot}2, 0)$, $s'=(15{\cdot}3, 4{\cdot}9)$, $\alpha=45^\circ$.

7. Through a given straight line $(ab, a'b')$ to construct a plane which makes a given angle α with a given plane pQr'.

Example: $a=(-12{\cdot}1, -5{\cdot}7)$, $a'=(-12{\cdot}1, 2{\cdot}8)$, $b=(3{\cdot}4, -1{\cdot}9)$, $b=(3{\cdot}4, +3{\cdot}8)$, $p=(7{\cdot}5, -26{\cdot}7)$, $Q=(23{\cdot}0, 0)$, $r'=(9{\cdot}2, 21{\cdot}0)$, $\alpha=22\tfrac{1}{2}^\circ$.

CHAPTER III

CURVED SURFACES AND SPACE-CURVES IN ORTHOGONAL PROJECTION

31. In analytical geometry it is usual to regard a surface in three dimensions as the locus of all points whose position in space is defined by some given law; the surface is the aggregate of all points in space which satisfy that law. Thus a circular cylinder is composed of all those points in space which are at a given distance from a given straight line. For the purposes of descriptive geometry, however, such a definition is evidently unsuitable; it is more convenient to regard a curved surface as generated by the motion of a curve in space, known as the *generating curve,* which changes its shape and its position in space according to some fixed law. The surface is thus the aggregate of all curves in space which were at any time coincident with the generating curve. Thus a circular cylinder may be regarded as generated by a straight line which moves in space in such a manner as always to be parallel to, and always at a given distance from, a given fixed straight line. As a rule, the generating curve is constrained always to rest on a given curve called the *directrix,* and thus its position in space is more or less determined. For example, a cylinder (generally) may be regarded as made up of all those straight lines (generating curves) which pass through a given curve in space (directrix) and are parallel to a given straight line. The directrix may therefore be regarded as any section whatever of the cylinder. A cone is similarly defined as generated by all those straight lines which pass through a given curve in space and also pass through a given fixed point. Those surfaces which are composed of rectilinear generators form an important class, of which the cone and cylinder are more particular cases, and are

known as *ruled surfaces.* Another important class of surfaces are the *surfaces of revolution.* They are generated by the rotation of a curve in space around a straight line known as the *axis* of revolution. Thus a sphere is the surface of revolution generated by the rotation of a circle about one of its diameters. Evidently any section of a surface of revolution made by a plane at right angles to the axis of revolution is a circle, which is known as a *parallel.* The sections made by planes passing through the axis of revolution are all equal to each other, no matter what be the position of the plane, and these curves are known as *meridians,* and the corresponding planes as *meridian planes.*

The advantage of this method of definition, from the point of view of descriptive geometry, now becomes clear, for instead of representing the surface by the projections of a certain number of selected points on it, it is now possible to represent it by the projections of a certain number of generating curves. If a point on the surface has to be dealt with, then the projections of the generator passing through that point are considered, a more elegant process than that of considering the projections of adjacent points chosen at random on the surface. A surface may often be generated in either of several distinct ways, and the problem under consideration will determine, in such cases, the choice of a method.

The first few problems in this chapter will be devoted to the consideration of tangent planes to curved surfaces. The tangent plane at any point of a curved surface may be defined as the plane containing the tangent lines at that point to all curves traced on the surface and passing through the point. It is thus determined if the tangent lines to any two such curves are known. The following *lemmas* will be required, viz.:—

α. The projection of the tangent to a curve at any point on it is the tangent to the projected curve at the corresponding point.

β. The tangent plane to any cone or cylinder, at any point on it, contains the generating line through that point, and touches the surface along the whole length of the generating line.

γ. The tangent plane to a surface of revolution is perpendicular to the meridian plane through the point of contact.

32. Problem XIX.—*To draw the tangent plane to any cylindrical surface at a given point on it.*

The cross section of the cylinder need not necessarily be circular. Take as generating curve a generator of the cylinder, and as directrix the curve in which the cylinder cuts the horizontal plane of projection. Suppose the generators to be parallel to the line (ab, $a'b'$), so that if the directrix and the line (ab, $a'b'$) are known the surface is uniquely determined. Let m be the horizontal projection of the point in the cylinder at which the tangent plane is to be drawn. The vertical pro-

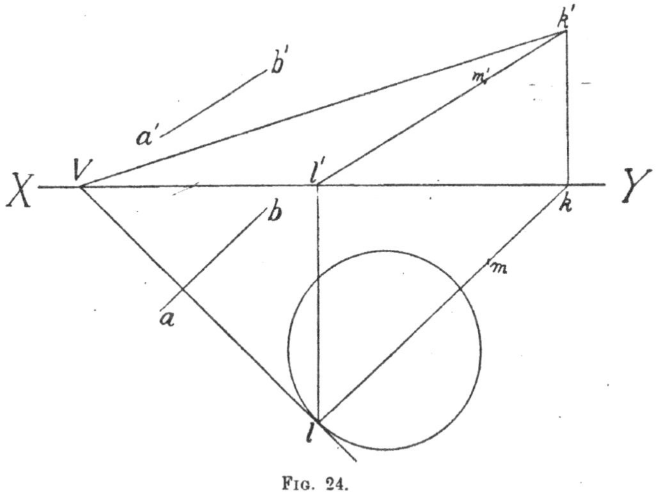

FIG. 24.

jection of the point need not be given, since the point is defined by its horizontal projection and by the fact that it lies on the given cylinder; part of the problem consists indeed in finding the vertical projection of this point from the given data. Now m necessarily lies on the horizontal projection of the generator through the point in question. This is a line parallel to ab passing through m and may therefore be constructed, suppose it meets the directrix in l. There may be more than one point l, which simply means that there are more points on the cylinder than one, whose horizontal projection is m. There will, however, be a finite, and, as a rule, a very limited number of such points to choose from, so that no real difficulty is introduced by this ambiguity. The point l is the horizontal trace of the generator, so that the vertical trace k' may be found by the usual trapezium construction, and $l'k'$ is the vertical projection of the generator on whose horizontal projection m lies. The vertical projection m' corresponding to m can now be found. The tangent plane at the given point passes through the generator $(lm, l'm')$, and it also passes through the tangent to the directrix l. Hence lV, which touches the directrix at l, is the horizontal trace of the required tangent plane. The corresponding vertical trace is the line through V and the vertical trace k' of the line $(lm, l'm')$.

Example: The directrix is a circle of 8 cm. radius and centre (15·7, −16·1). $a=(-13{\cdot}7, -17{\cdot}5)$, $a'=(-13{\cdot}7, 5{\cdot}9)$, $b=(0, -1{\cdot}8)$, $b'=(0, 14{\cdot}5)$, $m=(25{\cdot}6, -9{\cdot}1)$.

33. Problem XX.—*To draw the tangent plane at a given point to a cylindrical surface when the directrix is a curve in*

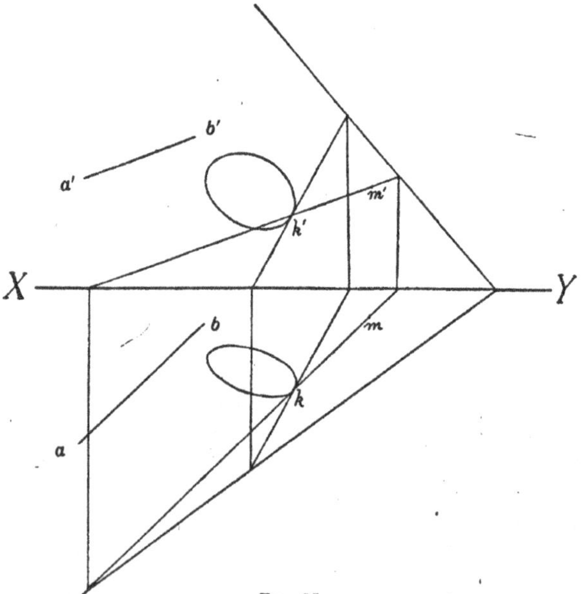

FIG. 25.

space whose horizontal and vertical projections are known, and the generators are parallel to a given line.

Let m be the horizontal projection of the point on the cylinder at which the tangent plane is required. Through m draw a line mk parallel to the horizontal projections of the generators, *i.e.* parallel to ab, so that mk is the horizontal projection of the generator through the point (m, m') on the cylinder. Let (k, k') be the point in which this generator meets the directrix, and through k' draw a line $k'm'$ parallel to the vertical projections of the generators, *i.e.* parallel to $a'b'$, so that $k'm'$ is the vertical projection of the generator considered. The tangent plane at (m, m') is the plane passing through the generator $(mk, m'k')$ and through the tangent to the directrix at the point (k, k').

This problem is a generalisation of the preceding one, and it is often simpler to reduce it to the less general case by first finding the intersection of the cylinder with the horizontal plane of projection and then proceeding as before.

> *Example:* The horizontal projection is a circle radius 5 cm., centre $(-12{\cdot}6, -8{\cdot}9)$, and the vertical projection a circle radius 5 cm., centre $(-12{\cdot}6, 15{\cdot}1)$, $m = (2{\cdot}4, -3{\cdot}6)$. Generators as in preceding example.

34. Problem XXI.—*To draw a tangent plane to a cylindrical surface through a point not on the surface.*

Suppose the cylinder to be defined by its base or curve of intersection with the horizontal plane of projection and by the direction of its generators. Since the given point is not on the surface it must be specified by both its horizontal and its vertical projections. Through the given point draw the line parallel to the generators and find its horizontal trace and from thence draw a tangent to the directrix. Then the required tangent plane contains the known line drawn parallel to the generators, and also the tangent line to the directrix. It may therefore be constructed.

> *Example:* The base is a circle of radius 7 cm. with its centre at the point $(0, -10)$. The vertical and horizontal projection of the generators make angles of 30° and 45° respectively with the ground line, and the cylinder as a whole is inclined to the right and away from the vertical plane of projection. The given point is $(-9{\cdot}5, -12)$, $(-9{\cdot}5, 5)$.

35. The methods of finding the tangent plane to a cone, and indeed to any developable surface whatever, very closely resemble the methods applicable to the cylinder which have just been outlined. Thus suppose it is required to draw a tangent plane to a given conical surface, at a given point on the surface. The method consists simply of constructing the generator, and then drawing at any point of that generator the tangent line to the directrix or section of the surface passing through this second point. The plane through the generator and this last line is the required tangent plane. The section of the surface chosen may be either the given directrix or it may be another section of the surface derived from it, as is most convenient. The other problems which arise may be treated in like manner.

> *Example:* A cone has a horizontal circular base of radius 7 cm., with its centre at the point (0, −8·5) and its vertex lies at the point (16·5, −5), (16·5, 18). To find the tangent plane at the point on the surface whose horizontal projection is (5·5, −7).

36. The preceding theory as regards the cylinder has an interesting application to the problem of finding the shortest distance between two straight lines *AB* and *CD*, that is to say, the length of a line drawn perpendicular to both of them. Through *AB* draw a plane parallel to *CD*, which can always be done by drawing through any point on *AB* a line parallel to *CD* and then through this new line and *AB* itself drawing the required plane. Now suppose that around *CD* as axis a circular cylinder be drawn the radius of whose principal section is the required shortest length. Then this cylindrical surface will touch the plane in a certain straight line which will cut the given line *AB* in a point *P*. If, now, through *P* there be drawn a line perpendicular to *AB* to cut *CD* in *Q*, then *PQ* will be the required shortest length. Hence the construction is as follows. Through the given line (*ab*, *a′b′*) draw a plane parallel to the second given line (*cd*, *c′d′*). Then from any convenient point of the line (*cd*, *c′d′*), such as one of its traces, draw a perpendicular on to the constructed plane. The foot of the perpendicular will be a point on the line of contact of the plane with the cylindrical surface. The line of contact (*mn*, *m′n′*) can now be drawn, for it is parallel to the line (*cd*, *c′d′*). This line will intersect the line (*ab*, *a′b′*) in (*p*, *p′*), from which point the perpendicular (*pq*, *p′q′*) to (*cd*, *c′d′*) can be drawn and its length, which is the length required, can be found.

> *Example:* Vertical trace of first line (5, 15), horizontal trace (−11·5, −19); vertical trace of second line (−6, 12·5), horizontal trace (10·5, 5).

37. Problem XXII.—*To draw the tangent plane at a given point on a given surface of revolution.*

Take the horizontal plane of projection to be perpendicular to the axis of the surface, so that the axis has for its horizontal

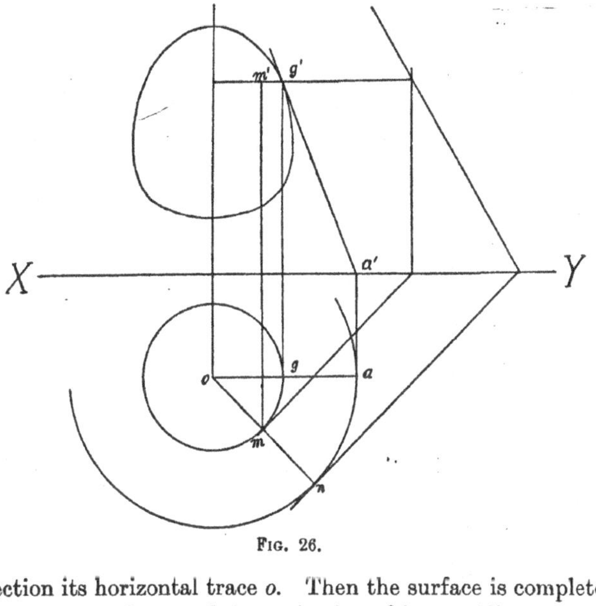

FIG. 26.

projection its horizontal trace *o*. Then the surface is completely defined by the point *o* and the projection of its meridian curve on the vertical plane.

Let *m* be the horizontal projection of the given point on the surface. Rabat the meridian plane in which *m* lies round the axis of the surface until it is parallel to the vertical plane of projection. Then if *g* is the new position of *m*, *mg* is a circular arc of centre o. Draw gg' to meet the principal meridian in g', the vertical projection corresponding to g. Now reverse the rabatting so that g' will move in a horizontal line, and if mm' be drawn parallel to the axis to meet this horizontal line in m', then m' will be the vertical projection corresponding to m.

The required tangent plane at (m, m') will pass through the tangent line to the meridian at (m, m'), and also through the tangent line to the parallel at (m, m'). To find the tangent line to the meridian, rabat (m, m') again into the position (g, g'), and draw the tangent line to the principal meridian at g'. This will have projections ga and $g'a'$, its horizontal trace being a. Now all tangent lines to the meridians at points on the same parallel evidently meet the horizontal plane of projection in points which are at the same distance from the axis. Hence if a circle of centre o and radius oa be drawn to meet om produced in n, n will be the horizontal trace of the tangent line to the meridian at the point (m, m'), and consequently this tangent line is known. The tangent at n to the circle of centre o will in fact be the horizontal trace of the required tangent plane. The tangent line to the parallel through (m, m') is a horizontal line whose horizontal projection is the tangent at m to the circle of centre o and radius om and whose vertical projection is the horizontal line through m'. Thus two lines in the required tangent plane are known, and consequently the tangent plane can be drawn.

> *Example:* The solid of revolution is a sphere of radius 6·5 cm. with its centre at the point (2·5, −12·5), (2·5, 10). The horizontal projection of the given point is (5·7, −9·2).

38. Problem XXIII.—*Through a given straight line to draw the tangent planes to a given sphere.*

The general method of procedure, which is also applicable to the cases of other solids than spheres, is as follows. With a chosen point on the given straight line as vertex, draw the tangent cone to the given sphere. Then the tangent planes to the cone which pass through the given line are the required tangent planes to the given sphere. By a suitable choice of the point at which the vertex of the tangent cone is to lie, the problem may be simplified and rendered amendable to graphical treatment.

Let (o, o') be the centre of the given sphere, and $(ab, a'b')$ the given straight line. Take as vertex of the tangent cone the point (a, a') in which $(ab, a'b')$ meets the horizontal plane through (o, o'). The plane in which the circle of contact of the

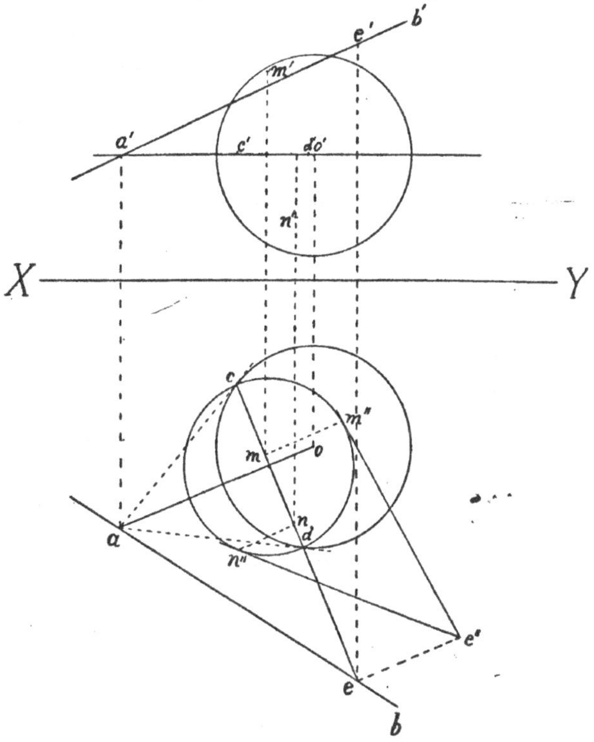

FIG. 27.

tangent cone lies will then be vertical, since it is perpendicular to the horizontal line $(oa, o'a')$; and if ac and ad be the tangents drawn from a to the circle o, cd will be the horizontal trace of this plane. Also if cd produced meets ab in e, then e will be the horizontal projection of the point in which $(ab, a'b')$ meets the plane of the circle of contact. Now rabat this plane about its horizontal trace cd so that it coincides with the horizontal plane of projection. The circle of contact of the tangent cone rabats into the circle on cd as diameter, and the point (e, e') into a point e'' on the perpendicular to cd through e, so that the distance ee'' equals the vertical height of e' above the horizontal line $a'o'$. From e'' draw $e''m''$ and $e''n''$ tangential to the rabatted circle cd, so that $e''m''$ and $e''n''$ are the rabatments of the tangents through (e, e') to the contact-curve, and the points m'' and n'' the rabatments of the points of contact of the required tangent planes. If (m, m') be the point of which m'' is the rabatment, m will be the foot of the perpendicular from m'' on cd, and the perpendicular distance of m' from $a'o'$ will be equal to the length mm''. The points m' and e' will be on the same or on opposite sides of $o'a'$ according as m'' and e'' are on the same or opposite sides of cd. Thus (m, m') and likewise (n, n') can be constructed, and the tangent planes through the given line can now be drawn. The problem is evidently soluble when, and only when, the point e'' lies on or without the circle on cd.

The problem may also be solved graphically by choosing as vertex of the tangent cone the point at infinity on the given straight line so that the tangent cone becomes a cylinder whose generators are parallel to the given line. The curve of contact is the great circle whose plane is perpendicular to the given straight line, and the points of contact of the required plane are the points at which the great circle is touched by the tangents drawn to it from the point of intersection of its plane with the given straight line.

Example: Radius of sphere 6·5 cm., centre at point $(0, -11)$, $(0, 13{\cdot}5)$, $a=(-12{\cdot}5, -14)$, $a'=(-12{\cdot}5, 8{\cdot}5)$, $b=(6, -24{\cdot}5)$, $b'=(6, 14)$.

39. Problem XXIV.—*Through a given point to draw the common tangent planes to two given spheres.*

Let (m, m') and (n, n') be the projections of the centres of the two given spheres, the spheres being represented by the projections of the great circles, whose planes are parallel to the planes of projection, and let (p, p') be the given point. If a right circular cone were circumscribed to the spheres so as to touch them both, the vertex of this cone would lie on the line joining the centres of the spheres. To find this vertex draw the external common tangents of the horizontal projections of the two spheres; then their point of intersection a will be the horizontal projection of the vertex in question, and the corresponding vertical projection can be deduced from it. Moreover, ap and $a'p'$ are evidently the projections of the line of intersection of two of the common tangent planes to the given spheres through (p, p'), so that these tangent planes may be constructed. The two other common tangent

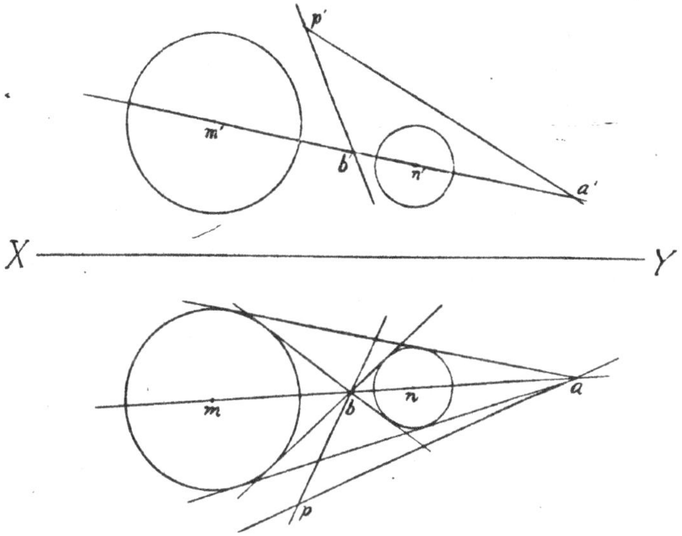

FIG. 28.

planes are obtained by finding the intersection b, the internal common tangents of the horizontal projections of the spheres, and proceeding as before.

40. The intersection of two surfaces S and S' is a curve in space, which will not in general be a plane curve. The problem thus arises—*given the specifications of two surfaces S and S', to find the projections of their curve of intersection* The general method adopted for the solution of the problem is as follows:—Take any horizontal plane P and find the horizontal projections of its curves of intersection with S and with S'.

Then the points of intersection of these projected curves will be the horizontal projections of points on the required space-curve. The corresponding vertical projections will lie on the vertical trace of the horizontal plane P, and can therefore easily be found. Thus one or more points in the curve of intersection is completely known, and by taking a new position of the plane P and repeating the process further points can be found. The curve of intersection can thus be constructed. Though in the general case the plane P is taken to be horizontal, in particular cases a different choice of this auxiliary plane would to a great extent simplify the problem. Thus, for example, in finding the intersection of two cylindrical surfaces, it would be most natural to take as auxiliary plane a plane parallel to the generators of both cylinders, as the plane would intersect the cylinders in straight lines, thus simplifying the construction to a very great extent.

It is often required to find the tangent line to the curve of intersection of two surfaces at any given point on the intersection. This tangent line arises as the intersection of the tangent planes to the two surfaces at the given point, and can be found in this way; but, as a rule, it is simpler to attack the problem directly by first constructing the projections of the curve of intersection and then drawing the respective tangents to these projections. This second method is particularly adapted to the cases in which one or both of the projections is a simple curve, such as a circle, ellipse, etc.

An example of the general method of finding the plane section of a curved surface will be given in the next section, and succeeding sections will deal with space-curves in general, and with the origin of twisted curves in space by the intersection of two curved surfaces.

41. Problem XXV.—*To find the section of a given cylinder of circular base made by a given plane.*

Take the vertical plane to be parallel to the generators and let the base of the cylinder rest on the horizontal plane. Then the horizontal projection of the cylinder lies in the region enclosed by the semicircle *fag* and the horizontal projection of the generators *ef* and *gh*, which are respectively the generators which are most remote from, and closest to, the vertical plane of projection. The vertical projection may similarly be represented by the vertical projections $a'b'$ and $c'd'$ respectively of the generators on the extreme left and the extreme right, as seen by an observer facing the vertical plane. These vertical projec-

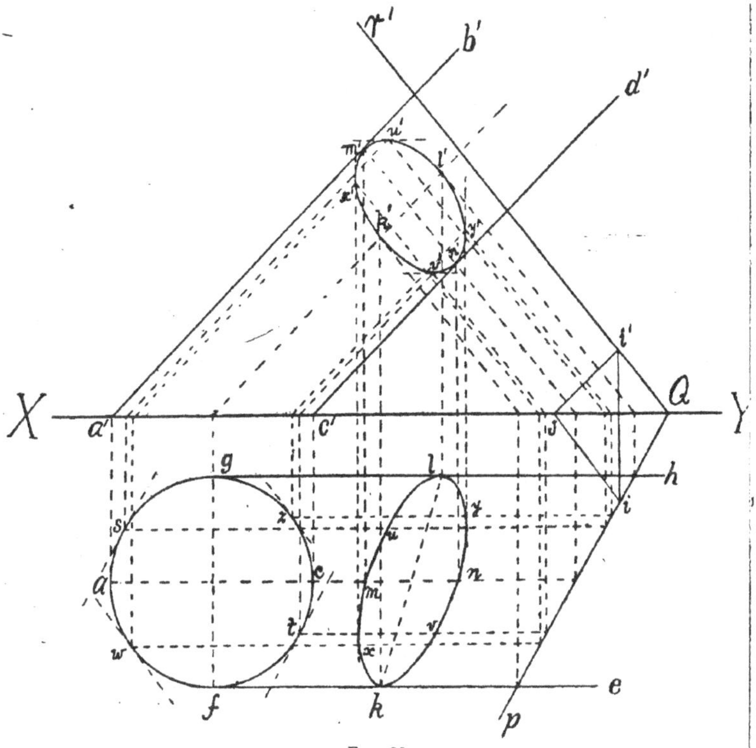

FIG. 29.

tions are evidently parallel to the actual generators themselves. The limits thus defined within which the projections lie will be called the *apparent contours* of the cylinder. Let pQr' be the given secant plane. Its sections with the cylinder will in no case lie outside the apparent contours. Find the points (k, k') and (l, l') in which the generators of which ef and gh are the horizontal projections meet the secant plane; k and l will evidently be on ef and gh respectively. In like manner find the points (m, m') and (n, n'), in which the generators of which $a'b'$ and $c'd'$ are the vertical projections intersect the secant plane. Now find the highest and the lowest points on the vertical projection of the curve. The tangents at the corresponding points on the actual section are horizontal, so that their points of contact are the intersections of the secant plane with the generators in which the cylinder is touched by the tangent planes parallel to the horizontal trace pQ. The construction is carried out by drawing the two tangents to the horizontal trace of the cylinder which are parallel to pQ. Let s and t be their points of contact, through which draw the generators through s and t cutting the secant plane in (u, u') and (v, v'). Then u' and v' are the highest and lowest points respectively on the vertical projection. Lastly, find the points on the two sections which are to the extreme left and to the extreme right, the tangents to which lie in the plane perpendicular to the ground line. For this purpose draw a plane at right angles to the ground line, with trace i, i', cutting pQ in i and Qr' in i'. At i' draw $i'J$ parallel to the generators of the cylinder, meeting the ground line in J. Then iJ will be parallel to the traces of the tangent planes at the required points. Draw the tangents to the base of the cylinder which are parallel to iJ, and let w and z be the points of contact. Then if the generators through w and z meet the secant plane in (x, x') and (y, y'), these last will be the two required points. Thus a certain number of distinctive points have been found on the projections of the curve of intersection which can now be drawn.

> *Example:* The radius of the base is 6·5 cm., its centre is at the point $(-8{\cdot}5, -8{\cdot}5)$. The generators are parallel to the vertical plane and make an angle of 45° with the horizontal plane. $p=(-14, 8)$, $Q=(15, 0)$, $r'=(0, 18)$.

42. A curve in space is completely represented, in Monge's Method, by its projections on the two planes of projection, and consequently may virtually be replaced by these two plane curves. This introduces a great simplification into problems connected with space-curves, for it is very much simpler to deal with plane curves than with twisted curves in space.

Given the projections of a curve in space, the first question that naturally arises is whether or not the curve so represented is plane. To settle this question, take any four points a, b, c, and d at random on the horizontal projection, and a', b', c', and d' the corresponding points on the vertical projection. Then join ab and cd and find their point of intersection e, and likewise join $a'b'$ and $c'd'$ and construct their point of intersection f'. Then a *necessary* condition that the curve should be plane is that the line ef' should stand at right angles to the ground line, so that if such is not the case, the curve is at once seen to be twisted. If, on the other hand, ef' does stand at right angles to the ground line, the probability is that the curve is plane, but in this case four other points should be taken and the construction repeated, and so on, until the true nature of the curve becomes evident.

43. Problem XXVI.—*At a given point on a curve to draw the tangent line, the osculating plane, the principal normal, and the binormal to the curve.*

If (p, p') be the given point, then the tangent at p to the horizontal projection of the given curve is the horizontal projection of the required tangent line, and similarly with the vertical projection. This construction may be carried out in the manner indicated in Section **7**.

The *osculating plane* intersects the given curve in three adjacent and ultimately coincident points, and consequently the tangent line at any point of the curve is contained in the osculating plane at that point. Thus since the tangent line at the given point has been drawn, its horizontal and vertical traces, which lie respectively on the horizontal and vertical traces of the required osculating plane, can be constructed, and it only remains to find one other point on the horizontal (or vertical) trace of the osculating plane in order that the latter may be determined. The horizontal trace can, however, be found directly as follows:—Let $(pq, p'q')$ be the tangent at (p, p') and let its horizontal trace be r. Now let $(p_1q_1, p_1'q_1')$ be a line drawn parallel to $(pq, p'q')$ through an adjacent point of the curve, and let r_1 be its horizontal trace. Then the locus of r_1 will be seen to be a curve having a cusp at r, and the tangent at r will be the horizontal trace of the required osculating plane, being ultimately the line joining the horizontal traces of two coincident tangent lines at p. The vertical trace can now also be drawn.

The *principal normal* to the curve at (p, p') lies in the corresponding osculating plane and is perpendicular to the corresponding tangent line, and thus it can easily be constructed by the methods of Chapter II. The *binormal* at (p, p') is perpendicular to the osculating plane through (p, p'), and can also readily be constructed.

44. Problem XXVII.—*To construct the curve of intersection of two given right circular cones whose axes are vertical, and*

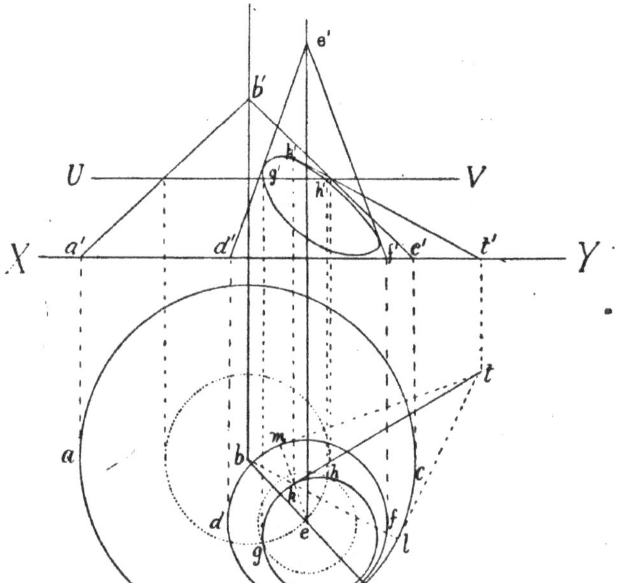

FIG. 30.

to draw the tangent line at any given point of the curve of intersection so constructed.

Let the vertical plane be taken parallel to the axes of the cones so that their vertical projections are the apparent contours

$a'b'c'$ and $d'e'f'$ respectively, b' and e' being the vertical projections of the corresponding vertices. The horizontal projections will be the two circles whose centres are b and e and represent the outlines of the bases of the cones. Now let UV be the vertical trace of a horizontal plane. It will cut the two cones in circles whose horizontal projections are concentric with the outlines of their respective bases and can be immediately drawn. These circles will intersect in two points g and h provided the plane UV is chosen so as to pass through the intersection of the cones, and these two points will lie on the horizontal projection of the curve of intersection. The corresponding vertical projections g' and h' will of course lie on UV. By changing the position of the plane UV further points on the curve of intersection may be constructed; its horizontal projection will be found to be a circle whose centre lies on be produced.

Now let (k, k') be the point on the curve of intersection at which the tangent line is to be drawn. Join bk and produce it to meet the circumference of the base of the cone $(abc, a'b'c')$ in l, and draw lt the tangent to the base at l. Then lt is the horizontal trace of the tangent plane touching this cone along the generator through (k, k'). Similarly, draw mt the corresponding horizontal trace of the tangent plane to the second cone through (k, k') meeting lt in t. Then t is the horizontal trace of the tangent line through (k, k'), and thus horizontal and vertical projections kt and $k't'$ of the tangent to the curve of intersection can be drawn.

Example: $b=(0, -12{\cdot}5)$, $b'=(0, 11{\cdot}5)$, $e=(4, -17)$, $e'=(4, 16)$; vertical angle at $b'=90°$, at $e'=22\frac{1}{2}°$.

45. Problem XXVIII.—*To construct the curve of intersection of two solids of revolution whose axes are coplanar.*

Choose the horizontal plane so as to be at right angles to the axis of one of the surfaces and the vertical plane so as to be parallel to the plane of the two axes. Let a be the horizontal trace of the axis of the first solid and $(ab, a'b')$ the projections of the axis of the second solid. The problem might be solved by the use of auxiliarly secant planes parallel to the vertical plane, but it is more convenient to consider the sections of the surfaces made by a family of spheres whose centre lies at the intersection of the two axes of revolution. Consider one of this family of spheres. It will cut the surfaces of revolution in circles whose planes are perpendicular to the respective axes of revolution, and therefore perpendicular also to the vertical plane of projection. Let $c'd'$ and $e'f'$ be the vertical traces of these circles, c', d', e', and f' being the points where the apparent contours of the surfaces is cut by the circle whose centre is a' and whose radius is the radius of the sphere considered. Let $c'd'$

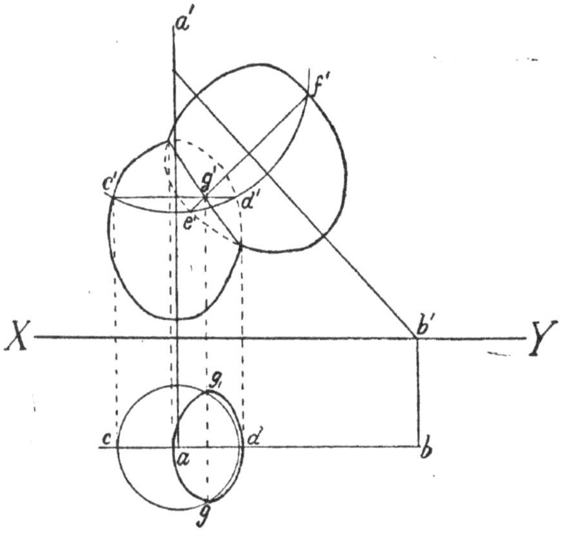

FIG. 31.

and $e'f'$ intersect in g'; then g' is a point on the vertical projection of the curve of intersection. To find the corresponding point (or points) on the horizontal projection, draw a circle with centre a and diameter equal to $c'd'$. This circle will evidently be the horizontal projection of the circle in which the sphere intersects the surface of revolution whose axis is vertical. The intersections g and g_1 of this circle, centre a, and a line drawn through g' perpendicular to the ground line will be two points on the horizontal projection of the curve of intersection. Further points may be obtained *ad libitum* and the curve of intersection gradually built up.

Example: The first solid is a circular cylinder the radius of whose cross-section is 5 cm.; the second solid is a circular cone of semi-vertical angle 30°, whose vertex lies on a generator of the cylinder and whose axis is inclined at a angle of 45° with the vertical.

46. Contour Lines.—The two concluding sections of the chapter will be devoted to the treatment of curved surfaces by means of the method of contours (§ **5**). This method is particularly applicable to what are called *topographical surfaces*, that is to say, surfaces which are defined by the equation $z=f(x, y)$ where $f(x, y)$ is any function whatsoever of its two arguments x and y. For if the (x, y) plane is taken as the horizontal plane of projection and the surface is cut by the parallel planes $z=0, 1, 2, 3 \ldots$ (that is to say, by planes whose *levels* are $0, 1, 2, 3, \ldots$, the z-axis being vertical), the resulting curves or *contour lines* $f(x, y)=0, 1, 2, 3 \ldots$ projected orthogonally on to the (x, y) plane give a representation more or less complete according to the smallness or largeness of the unit of length chosen. With any given scale, the closer the contours are crowded together at any given point, the steeper will be the surface at the corresponding point on it.

The practical utility of this method lies in its application to such surfaces as cannot be defined by an *analytic* function $f(x, y)$, and are of such irregularity that they cannot be classified with any surfaces known to pure geometry. An example, and indeed the principal example of such a case, is the surface of the earth, and the application of the above methods leads to the science of topography. The method of contours is best applicable to those surfaces, such as the surface of the earth, which are cut by a vertical line in only one point, for in such cases the possibility of the overlapping of contours does not enter.

The chief problems that arise in topographical projection are to determine the curve of intersection of two given surfaces, or of a plane and a given surface. This amounts simply to finding the points of intersection of the two respective contours at the same level and thus gradually building up the required curve of intersection. In actual practice the most important problem is to find the *profile* or section made by a vertical plane, which is a particular case of the above.

47. Problem XXIX.—*To find the curve of intersection of a given topographical surface made by a given plane.*

The surface is defined by its contours at levels 0, 1, 2, 3, . . . and the plane by its line of steepest slope. Through the points 0, 1, 2, 3, . . . on the latter draw the horizontal lines (which will be perpendicular to the line of steepest slope) to cut the corresponding contours in one or more points. Then the curve joining successive points of intersection will be the required section.

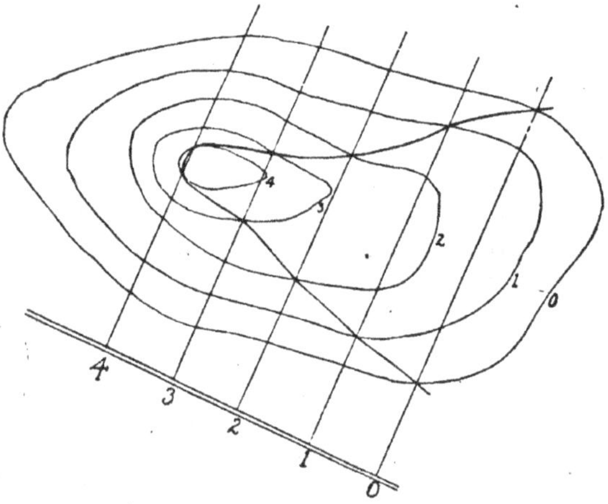

FIG. 32.

Example: A topographical map is composed of eleven circles whose centres lie on a given horizontal line 40 cm. long. The first circle has its centre on the mid-point of the line, and its radius 20 cm. The other circles have their centres 1, 2, 3, . . . 10 cm. to the right of the mid-point of the line, and their radii 18, 16, 14, . . . cm., the last circle being a mere point. The levels of the contours are respectively 0, 1, 2, . . . 10. It is required to construct the curve of intersection of the surface made by a plane passing through its highest point, the projection of its line of steepest slope making an angle of 30° with the given horizontal line, the interval being 2·5 cm.

EXERCISES ON CHAPTER III

1. Given two intersecting straight lines $(ab, a'b')$, $(ac, a'c')$, to construct a plane passing through the first and making a given angle α with the second.

Example: $a=(-12{\cdot}5, -7)$, $a'=(-12{\cdot}5, 9{\cdot}5)$, $b=(8, -10{\cdot}5)$, $b'=(8, 14)$, $c=(11, -6{\cdot}5)$, $c'=(11, 9)$, $\alpha=30°$.

2. To construct the common normals to a given cone and a given cylinder.

Example: The cylinder is vertical and has a circular base of radius 6 cm. in the horizontal plane. The cone has a circular base of radius 8 cm., also in the horizontal plane, its centre being 27 cm. distant from that of the base of the cylinder. The vertex of the cone is 15 cm. above the horizontal plane and at a distance of 9 cm. from the axis of the cylinder. The horizontal projection of the line joining the vertex of the cone to the axis of the cylinder makes an angle of 45° with the line joining the centres of the bases.

3. A point (o, o'), situated at a distance of 10 cm. from each of the planes of projection, is the common centre of a sphere S of radius 4 cm. and of a horizontal circle c of radius 2 cm. In the horizontal plane the point o is the centre of a circle C of radius 8 cm. A straight line $(ab, a'b')$ starts from a point in the circumference of the circle C and touches the circle c in such a manner that its horizontal projection is tangential to that of the circle (this line is not to be taken parallel to the vertical plane).

It is required (i.) to trace the hole that would be made in the sphere by a circular cylinder having $(ab, a'b')$ as axis and radius 2 cm., and (ii.) to find the section of the sphere with the circular hole in it, made by a vertical plane passing through the axis of the cylinder. (École polytechnique, 1853.)

4. A plane pQr' is given whose horizontal trace pQ makes an angle of 30° with XY, the plane being inclined at an angle of 53° to the vertical plane. A point (a, a') in this plane is situated at a distance of 2 cm. from the horizontal and 12 cm. from the vertical plane. This point is the centre of the base of a right cone seated on the upper face of the plane and whose radius is 3 cm. The height of the cone is 12 cm. To construct the projections of the cone. (École navale, 1876.)

5. A sphere of radius 50 mm. touches the horizontal plane at the point o. From a point a in the horizontal plane such that $oa=100$ mm. are drawn in succession (i.) the two tangents to the sphere which make, with the vertical, an angle of 40°, and (ii.) the two tangents to the sphere which make, with the horizontal line oa, an angle of 23°. These four tangents being considered as the edges of an indefinite pyramid of vertex a, to trace the projections of the volume common to the sphere and the pyramid. (École navale, 1895.)

CHAPTER IV

PERSPECTIVE

48. Perspective Representation of Regular Solids, etc.—The representation in perspective (§ **6**) of such bodies as cubes, rectangular blocks, and so on, which are bounded by plane faces, is an important section of this branch of descriptive geometry. The edges of such solids form a number of sets of parallel lines, and as each set of parallel lines has the same vanishing point, these vanishing points will certainly play a considerable part in the solution of the problem in question. The vanishing line of the ground plane, and therefore of all horizontal planes, is a horizontal line through the centre of vision, which is often termed the *horizon.* It contains the vanishing points of all horizontal lines, and when the orientation of these lines are known, their vanishing points are fully determined.* The vanishing point of a non-horizontal system of parallel straight lines lies vertically above the vanishing point of their orthogonal projections on any horizontal plane, the latter vanishing point lying of course on the horizon. If the angle which the system makes with the horizontal is known, the vanishing point can be completely determined. When the vanishing points of all the systems of parallel lines have been constructed, it only remains to utilise the data of the problem to determine the apparent position of the lines of each system and their apparent magnitude. The manner in which this is carried out is best explained by means of an example.

* The vanishing points of horizontal lines making an angle of 45° with the picture plane are called the *distance points.* Their distance from the centre of vision is plainly equal to the distance of the latter from the point of sight.

49. Problem XXX.—*To construct the perspective representation of a given cube.*

Suppose that the cube lies on the ground plane, with one of its vertical edges touching the picture plane, and its sides making given angles α and β with the picture plane. Let XY be the ground line or intersection of the picture plane with the ground plane and let C_0 be the centre of vision. Now suppose that the horizontal plane through C_0 is rabatted about the

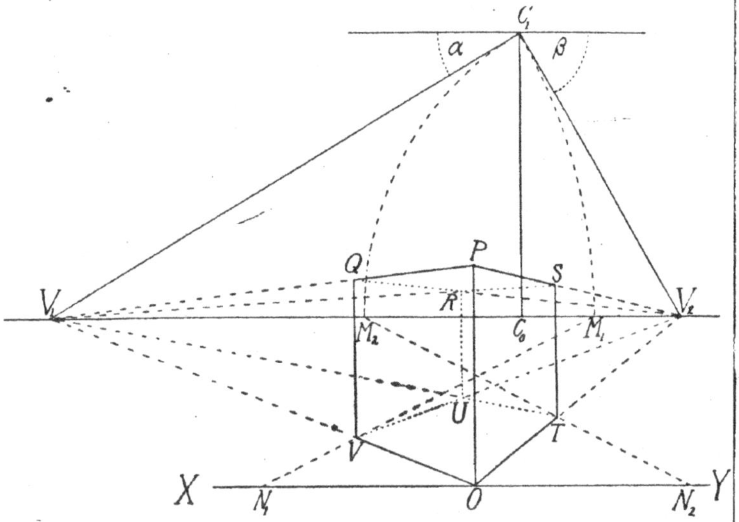

FIG. 33.

horizon line until it coincides with the picture plane. The rabatted position C_1 of the point of sight will be vertically above C_0 and the distance C_0C_1 will be known. Through C_1 draw the lines C_1V_1, C_1V_2, making angles α and β with the horizontal. These lines will be the rabatted positions of the straight lines through the point of sight parallel to the horizontal edges of the cube. Hence the points V_1 and V_2 in which they intersect the horizon will be the corresponding vanishing points. Let O (a point on the ground line) be the foot of the edge which touches the picture plane. Draw OP perpendicular to XY with its length equal to the length of an edge of the given cube, so that OP represents in magnitude and position the edge which touches the picture plane. Now join V_1O and V_1P, which will represent the direction and position of two opposite sides of a face of the cube.

The next step is to mark off on OV_1 and PV_1 the apparent lengths of the corresponding edges of the cube. Now the lengths of the perspective representations of lines whose true length in space is known are found by means of what are called *measuring points*. The measuring point of a system of parallel lines, such as these edges, is defined to be a point M_1 in the picture plane, in the same horizontal line as the vanishing point V_1, and such that V_1M_1 is equal in length to V_1C, where C is the station point; so that in the diagram V_1M_1 is equal to V_1C_1. If, now, OA be the actual lower edge in space and OV its representation on the picture plane, suppose the (plane) figure $COVA$ parallel-projected on to the picture plane by lines parallel to CM_1. The point C is projected into M_1, so the space line CV_1 is projected into M_1V_1 which is equal to it in length; and as OA is parallel to CV_1, OA will be projected into a segment ON_1 parallel to M_1V_1 and of length equal to OA.

From this it will be evident that the construction required is to mark off on the horizon a length V_1M_1 equal to V_1C_1, and on the ground line a length ON_1 equal to the true length of the edge, and to join M_1N_1, cutting OV_1 in OV; then OV will be the complete representation of one of the edges of the cube which passes through O.

Since vertical lines have no finite vanishing point, they remain vertical when projected on to a vertical picture plane. Hence the vertical edge through V can immediately be constructed, its upper extremity Q lying on PV_1. Thus a complete face of the cube $OPQV$ has been constructed, and similarly the adjacent face $OPST$ can be built up, so that the projections of six corners of the cube are now known. If V_1S and V_2Q intersect in R and V_1T and V_2V in U, R and U will be the two remaining corners, so that the cube has been completely represented.

In the above problem, one of the edges of the cube was in contact with the picture plane, and thus appeared in its true dimensions and position. If the given figure does not touch or intersect the picture plane, a point from which to begin the construction can always be obtained by producing one or other of the edges until it meets the picture plane.

> *Example:* A rectangular block height 10 cm. rests on the ground. Its base is 8 cm. square and the nearest edge touches the picture plane. The vertical edges make angles of 60° (on the left) and 30° (on the right) with the picture plane. Construct its perspective representation as seen from a station point 12 cm. in front of the picture plane, 6 cm. above the ground plane, and 3 cm. to the left of the vertical plane through the nearest edge of the cube perpendicular to the picture plane.

50. Solution of Perspective Problems by Monge's Method.—Perspective representations of objects in space can also be obtained directly by the methods of ordinary descriptive geometry. The picture plane on which the perspective is to be

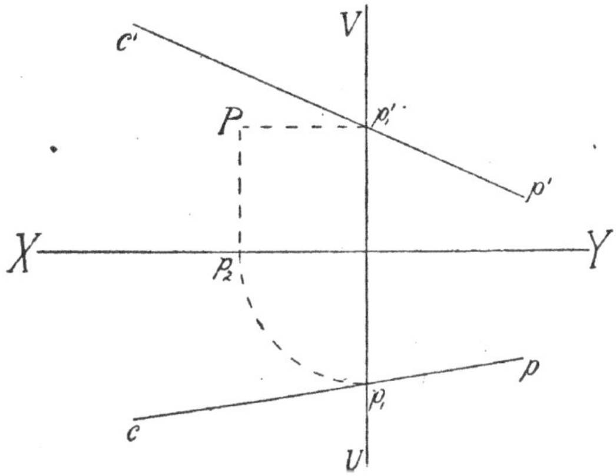

FIG. 34.

formed is taken in the first place to be perpendicular to the ground line XY, so that its traces form a continuous line UV. Let (c, c') be the plan and elevation of the station point, and let (p, p') be the point whose perspective representation is desired. Then if the straight line cp intersects UV in p_1, this point will be the plan of the perspective representation of the point (p, p'). Similarly, p_1', the intersection of $c'p'$ with UV, will represent the corresponding elevation. Now let the picture plane be rabatted about UV into a position of coincidence with the plane of the paper. During the process p_1' will move along a line parallel to the ground line, and p_1 will move along the circumference of a circle whose centre is at the intersection of XY and UV, and will ultimately take up a position p_2. The plan and elevation of the perspective representation of (p, p') in the plane of the paper are now known, so that the corresponding point P may be constructed.

By reversing this process it is possible to proceed from the perspective representation of a point or set of points to the representation in orthogonal projection, and hence any problem in projection may be solved by Monge's Method. A more general way of attacking the same problem is given in Section **56**.

51. Perspective Representation of Plane Curves.—A plane curve is, in general, completely determined if its projection from a given station point on to a given picture plane, and the trace and vanishing line of the plane in which it lies, are known. This method of representing a plane curve in perspective is of considerable interest as giving rise to an elegant means of finding plane sections of conical and cylindrical figures, depending on Desargue's Theorem. This theorem states that if two figures, such as the given plane curve and its projection, are in perspective, then the points of intersection of corresponding straight lines in the two figures are collinear, and conversely. In the present case, a straight line joining any two points of the given curve meets the straight line joining the corresponding points of the projected figure on the intersection of the planes containing the given curves. In consequence of this the figures will always remain in perspective if either plane be rotated about its trace on the other, or *axis of perspective*, as it is sometimes termed. Hence, if the plane of the given curve be rabatted on to the plane of projection, the rabatted curve and its projection will still remain in perspective, the centre of perspective being the rabatted centre of projection.

The first step towards the solution of any problem which depended on the rabatting of the given curve on to the plane of projection would be to find the position of the rabatted centre of projection. During the rabatting the centre of projection* will evidently move in a plane perpendicular to the axis of perspective, and will finally lie in the trace of this plane. Its path, during rabatting, will be a circle whose centre lies on the vanishing line of the plane of the curve, so that its final position can readily be constructed. The next step in the construction consists in constructing the rabatted positions of points whose projections are given. If the axis of perspective LM and the rabatted position C' of the centre of projection are known, and if A' be the projection of a point whose rabatted position is known to be A, then if B' be the projection of any other point, its rabatted position B can be found. For since A and A' and likewise B and B' are corresponding points, AB and $A'B'$ must intersect upon LM, and BB' must pass through C', so that B may be constructed.

52. Problem XXXI.—*To find the section of a given right circular cone made by a plane inclined at a given angle to its base and passing through the edge of the base.*

The required section may be considered as a plane curve in space whose projection is the base of the cone, so that the vertex

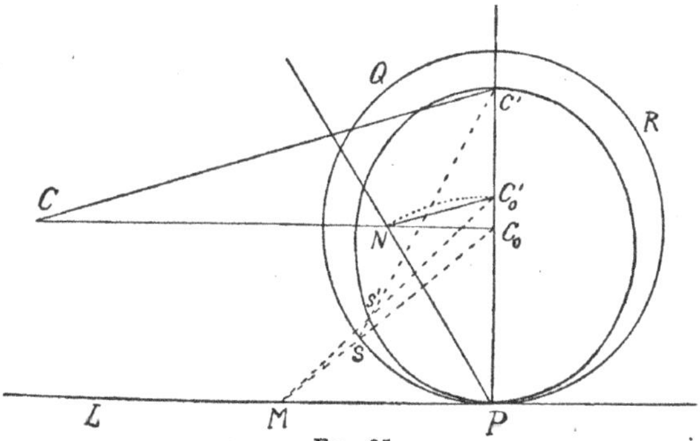

FIG. 35.

of the cone is the centre of projection. By the general method that has been outlined the plane of the section may be rabatted on to the plane of the base and hence the true nature and size of the section may be found.

* The centre of projection is now taken to mean the point in which the straight lines joining points and their projections are concurrent.

Let the circle PQR, centre C_0, be the base of the given cone. The tangent at P will be the trace of the secant plane and hence also the axis of perspective, P being the point in which this plane meets the base of the cone. Let C_0C be drawn through C_0 perpendicular to C_0P and equal in length to the height of the given cone, and let PN be drawn through P so that the angle C_0PN equals the angle which the secant plane makes with the base of the cone, and let PN meet C_0C in N. Along PC_0 mark off PC_0' equal to PN, and through C draw CC' parallel to NC_0', meeting PC_0 produced in C'. Then when the secant plane is rabatted about the axis of perspective so as to coincide with the base of the cone, C' will be the new position of the centre of perspective,* and, since PC_0' or PN is equal in length to the original line in space corresponding to PC_0, C_0 and C_0' will be a pair of corresponding points. Thus the rabatted centre of projection C' and two corresponding points C_0 and C_0' are known, so that the required section may be constructed. For if S be any point on the circle PQR, and if C_0S be joined and produced to meet the axis of projection in M, the intersection S' of $C'S$ and $C_0'M$ will be a point on the required curve, and a sufficient number of the positions of S' being known, the curve can be drawn. In the present case, the curve of intersection is known to be a conic, so that five separate positions of S' are sufficient for its complete specification.

If the given solid had been a cylinder and not a cone, the rabatted position of the centre of projection would have been at infinity, that is to say, C' would have been the point at infinity on the line PC_0, so that the corresponding points S and S' would have been on a line parallel to C_0P. Otherwise the construction is identical with the above.

Example: Height of cone 20 cm., radius of base 6 cm., angle between secant plane and base, $22\frac{1}{2}°$.

53. Representation of Twisted Curves and Curved Surfaces. —In order to represent a given curve in space by the method

* For if V denote the intersection of PC_0 with the vanishing line in the base, the centre of projection during the rabatting describes a circle round V (§ 51), so that CC' makes equal angles with VC and VC'; but (by the definition of the vanishing line) VC is parallel to the secant plane, so that CC' must make equal angles with the secant plane and the plane of the base, *i.e.* with PN and PC_0.

of perspective, it is sufficient to assign to each of its points the corresponding perspective description (TI', P'). The projection P' of any given point P is fixed if the centre and plane of projection are fixed, and hence P' will trace out a certain definite curve on the plane of projection. The auxiliary line (TI') is, however, to a considerable extent arbitrary, and the representation may be greatly simplified by imposing convenient restrictions on this auxiliary line. The most obvious of such restrictions are (i.) to regard the trace T as fixed so that all the lines TP pass through the same point T and hence trace out a cone, and (ii.) to regard the vanishing point I' as a fixed point so that all the lines TP have the same vanishing line and hence trace out a cylinder.

In the first case (fig. 36), the diagram in the picture plane will consist of the fixed point T and the curves traced out by

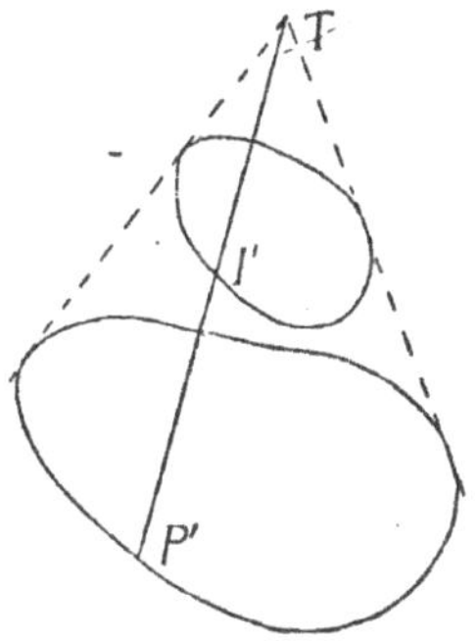

FIG. 36.

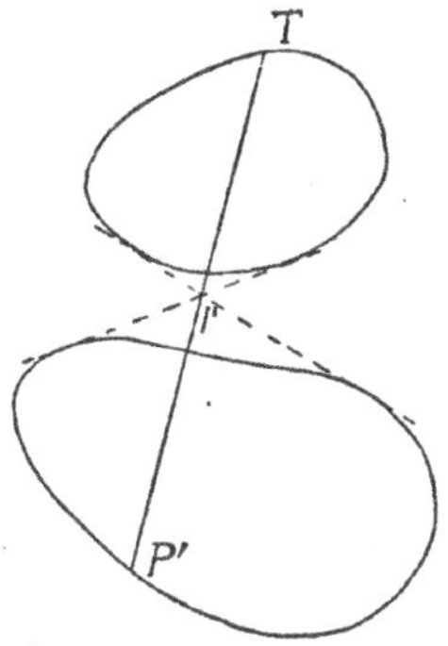

FIG. 37.

the moving points P' and I', and in the second case (fig. 37) it consists of the fixed point I' and the curves traced out by the moving points P' and T. If the space-curve is the same in the two cases, the path of P' will be the same in both, but the paths of T and I' will not be identical.

The tangent line at a point P on the given curve can readily be constructed, for its projection is simply the tangent line at P' to the projected curve. Now suppose the curve be represented in the first of the above modes. The curve in space, together with the auxiliary lines through T, form a cone whose vertex lies at T, and the tangent line to the curve at P will lie in that tangent plane to the cone which touches it along the generator TP. The vanishing line of this plane will touch the locus of I' at the point I' corresponding to P', and its trace will pass through T. Thus the trace and vanishing point of the required tangent line are also known. If the curve be represented in the second mode, it can be dealt with in an analogous manner.

As in the case of orthogonal projection, a solid is most conveniently represented by its director curve and generators. If these are known its apparent contour with respect to the centre of projection can then be found. Thus a cone is naturally specified by its vertex and its base (a plane curve) being given, so that its generating lines, and in particular those which define its apparent contour, may immediately be constructed. A cylinder is likewise specified by its base and, since its generators are parallel, the common vanishing point of its generators. When a curved surface is of unlimited extent, its *vanishing curve* or locus of points, which correspond to the points at infinity on the surface, will naturally be given.

54. Problem XXXII.—*From a given point to draw a tangent plane to a given cylinder.*

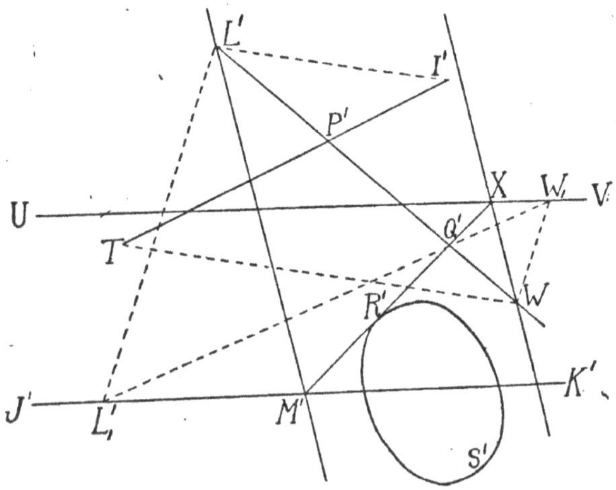

FIG. 38.

Let (TI', P') be the given point, UV the trace, and $J'K'$ the vanishing line of the plane of the base of the cylinder, and S'

the projection of the base itself, and let L' be the vanishing point of the generators of the cylinder. Through the point (TI', P') draw a line parallel to the generators; its vanishing point will be L' and its trace W will be such that TW is parallel to $I'L'$, and L', P', and W are collinear. Now find the projection Q' of the point in which this line meets the plane of the base. For this purpose draw any plane $(WW_1, L'L_1')$ through the given line, intersecting the plane of the base in the line (W_1L_1'), then this line will intersect the line (WL') in the required point Q'. If now $Q'R'$ be a tangent drawn from Q' to the curve S', it will be the projection of a tangent drawn to the base from a point in its own plane. The required tangent plane passes through this tangent line and is parallel to the generators of the cylinder. Thus its vanishing line will pass through L' the vanishing point of the generators and through M' the vanishing point of the tangent line whose projection is $Q'R'$. The corresponding trace will pass through the point X, the trace of this tangent line, and through W, the trace of the line through the given point parallel to the generators. Thus $(WX, L'M')$ is the required tangent plane.

55. Relation between the Method of Perspective and Monge's Method.—As either of these two methods furnishes a complete representation of the geometrical figures in question, it necessarily follows that if the representation according to one method is given, the corresponding representation according to the other method can, in general, be immediately deduced.

Thus suppose that (TI') is the perspective representation of a given line in space with respect to a given centre of projection C, and that it is required to construct its representation according to Monge's Method. For simplicity, take the plane of projection in the perspective system to be the horizontal plane of projection in the new system, and let XY, any straight line in this plane, be the ground line.

C_0, the foot of the perpendicular drawn from C on to the plane of projection, will become c the horizontal projection of C; and the corresponding vertical projection c' is known, for its distance above the ground line is equal to the distance of C from the original plane of projection. T, the trace of the line (TI'), is also the horizontal trace of the same line in orthogonal projection, and may be denoted by t; the corresponding vertical projection t' lies on the ground line, and can therefore be constructed.

Now consider the line CI' in space; it is parallel to the given line and hence its orthogonal projections pass through c and c' respectively, and are parallel to the projections of the given line. Its horizontal trace is I' (or i), and hence its horizontal projection is ci, and its vertical projection $c'i'$ can immediately be deduced. Hence the required horizontal and vertical projections are respectively the line through t parallel to ci and the line through t' parallel to $c'i'$. Suppose now that P is a given point on the line (TI') whose projection with respect to C is P', and that its

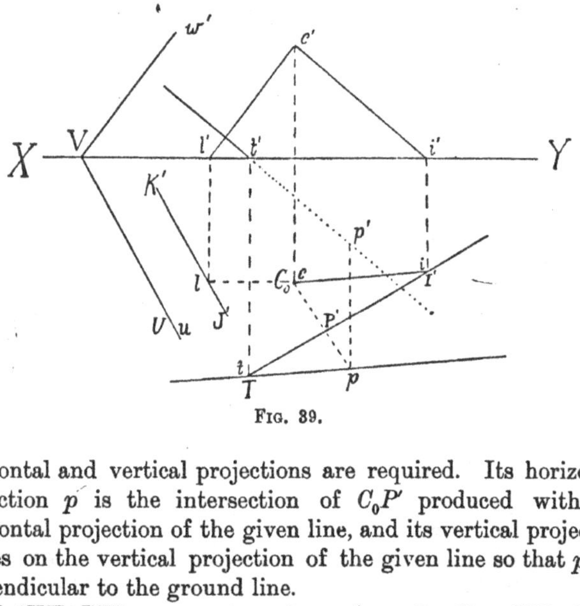

FIG. 39.

horizontal and vertical projections are required. Its horizontal projection p is the intersection of C_0P' produced with the horizontal projection of the given line, and its vertical projection p' lies on the vertical projection of the given line so that pp' is perpendicular to the ground line.

If $(UV, J'K')$ represent a given plane, the line UV will be its horizontal trace and may be denoted by uV, where V is taken to be on the ground line. Now consider an auxiliary plane through C parallel to the vertical plane of projection. Its horizontal trace will pass through c and will meet the vanishing line $J'K'$ in some point l. The plane $CJ'K'$ will meet the auxiliary plane in a line whose vertical projection is $c'l'$, so that $c'l'$ is parallel to the vertical trace of the given plane.

Hence if Vw' be drawn through V, parallel to $c'l'$, Vw' will be the vertical trace of the given plane.

If the horizontal and vertical projection of a figure in space are given, its projection on the horizontal plane from a given centre (c, c') may be found by a mere reversal of the steps indicated in the preceding examples. Thus if a straight line $(tp, t'p')$ be given, its trace T will be its horizontal trace, and its vanishing point I' will be the trace of the line drawn through (c, c') parallel to $(tp, t'p')$. If a point (p, p') be given, its projection P' will be the horizontal trace of the line $(cp, c'p')$, and if a plane uVw' be given its trace will be uV, and its vanishing line will be the trace of a parallel plane passing through (c, c').

EXERCISES ON CHAPTER IV

1. Put into perspective a rectangular block 8 cm. × 15 cm. × 12 cm. which rests on the ground. The nearest edge is 5 cm. to the right of the vertical plane and 8 cm. behind the picture plane. The longer edges are inclined at an angle of 30° to the picture plane and vanish towards the left. The station point is 6 cm. above the ground plane and 5 inches in front of the picture plane.

2. The above rectangular block is tilted about the left-hand longer side of its base until its base makes an angle of 15° with the ground plane. Construct its perspective representation in this position, the station point being as before. [Note that if V_3 be the vanishing point of a set of parallel lines inclined to the horizontal, and if V_1 be the vanishing point of their horizontal projections, V_3 lies vertically above V_1, and the angle $V_1M_1V_3$ will be the angle between a line of the system and its horizontal projection.]

3. An oblique pyramid of 25 cm. vertical height stands on a regular hexagonal base the length of whose sides is 11 cm. Its vertex stands vertically above one of the corners of its base. Find the section made by a plane passing through the base line of one of its two faces of least slope and making an angle of 45° with the plane of the base.

4. An oblique prism has as its base a parallelogram whose sides are 12 cm. and 9 cm. and whose angle is 60°. The edges of the prism make an angle of 45° with the plane of its base and are at right angles to the longer sides of the base. Construct the section of the prism made by a plane passing through one of the shorter sides of the base and making an angle of 30° with the plane of the base.

CHAPTER V

PHOTOGRAMMETRY

56. The Problem of Photogrammetry.—The art of *Photogrammetry*, or *Metrophotography* as it was originally termed by its inventor Laussedat (1851), has for its aim the obtaining from ordinary photographs correct metrical representations of the object photographed. Thus it is of considerable utility in military science, for it supplies a means of obtaining drawings of enemy fortifications from a few hastily taken snapshots. This was in fact the origin from which the art of photogrammetry sprang. It also finds application in the more peaceful occupation of research in the upper atmosphere, for the only means by which it is reasonably possible to secure accurate measurements of clouds is by direct photography followed by the application of the methods of photogrammetry.

The image which is obtained in the camera during the ordinary processes of photography is to all intents and purposes a perspective representation of the object to be photographed, the optical centre of the lens being the vertex of projection,* and the plane of projection being the sensitive plate which is placed in the focal plane of the lens. Thus the developed *negative* represents an image in the focal plane of the lens in perspective with the object photographed. The *positive*, or print taken from the negative, is evidently the image of this perspective figure, with respect to the centre of the lens, on a plane situated as far in front of the lens as the focal plane is behind it. Thus the positive image is equally a perspective representation of the

* Here the optical system is considered as a single thin lens, but Gauss' theory of cardinal points furnishes an adequate extension to the more general case of a compound system of lenses.

object in question, and has the additional advantage that it represents the object in its true aspect, and not, like the negative, as a mirror-image. In everyday language this positive image of the object is called its photograph, but would be more correct to use the term *photogram*, and this term is generally employed in connection with photogrammetry.* In practice the photograms are obtained, if possible, with an instrument known as the *photo-theodolite*, which is virtually a combined camera and theodolite. By its use not only is the photograph taken but at the same time the elevation and aspect of the centre of the field with regard to the station is observed. The problem of photogrammetry then resolves itself into the following:—Given two photograms of the same set of space points, the stations at which the camera is placed being supposed known, to determine the true relative positions of those space points. The stations at which the observations are taken may not necessarily be at the same level, and the central axes of the cameras may be inclined at any angle to each other and to the horizontal.

57. Solution of the Problem.—The first stage in the solution consists in finding the horizontal projection of the set of space

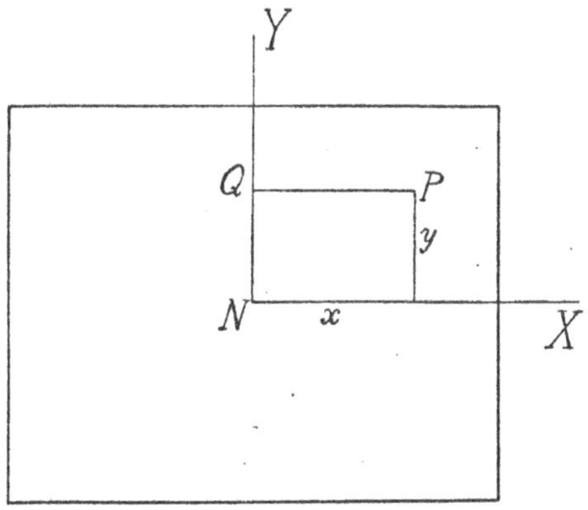

FIG. 40.

* The terms *telegraph*, *spectrograph*, *thermograph*, and so on, signify the instruments used for certain purposes, and the corresponding words *telegram*, *spectrogram*, *thermogram*, etc., the material results obtained by their use. Thus for consistency, the term *photograph* should signify the photographic camera, and the word *photogram* what is now signified by the term photograph.

points, the procedure being as follows. Let N be the centre of one of the photograms, that is to say, the point in which the sensitive plate is intersected by the central axis of the camera. Then the positions of points on the photogram are determined by measuring their rectangular co-ordinates (x, y) with respect to N, the x axis being horizontal and the y axis being the line of greatest slope in the plate. In refined practical work these measures are carried out in a specially constructed measuring machine, and can be made with precision.

Now let o_1 and o_2 be the projections on some convenient horizontal plane of the two positions of the optical centre of the

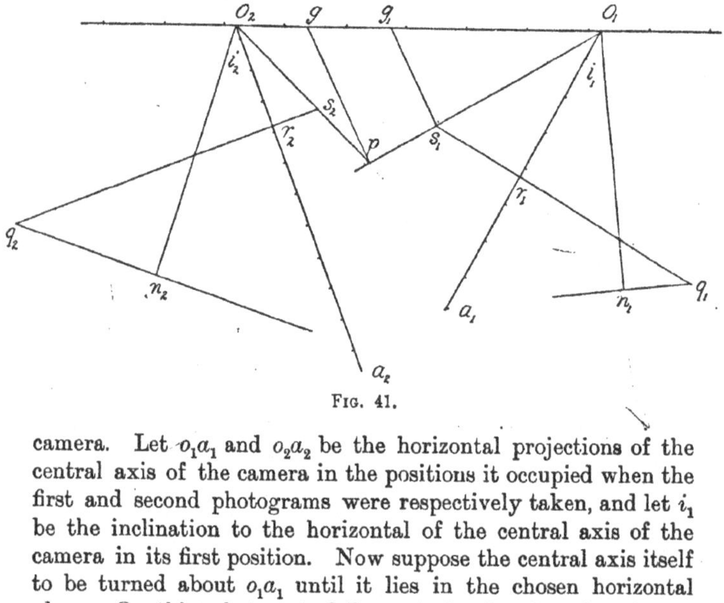

FIG. 41.

camera. Let o_1a_1 and o_2a_2 be the horizontal projections of the central axis of the camera in the positions it occupied when the first and second photograms were respectively taken, and let i_1 be the inclination to the horizontal of the central axis of the camera in its first position. Now suppose the central axis itself to be turned about o_1a_1 until it lies in the chosen horizontal plane. On this rabatment of the central axis take a length o_1n_1 to represent the focal length of the camera, so that a line through n_1, perpendicular to o_1n_1, will represent the rabatment of the axis of y in the first photogram. On this line take a point q_1 such that the length n_1q_1 is equal to y, the y-co-ordinate of a point P_1 in this photogram, and q_1 will be the rabatment of Q_1. Draw q_1r_1 perpendicular to o_1a_1: it represents the perpendicular by which Q_1 is projected orthogonally on to the horizontal plane, so that r_1 is the true horizontal projection of the point Q_1. Lastly, produce q_1r_1 to s_1 so that the length of r_1s_1 is x, the x co-ordinate of P_1; then s_1 will be the horizontal projection of P_1. Now repeat the whole construction in the case of the second photogram, obtaining in succession the points n_2, q_2, r_2, and finally s_2, the projection of the point P_2 corresponding to P_1 in the first photogram. Thus the points s_1 and s_2 which have been obtained are the horizontal projections of the two images of the same space point P; and therefore o_1s_1 and o_2s_2 are the horizontal projections of the line of sight from the optical centre of the camera in its first and second positions respectively to the actual space point. Hence if o_1s_1 and o_2s_2 intersect in p, then p will be the horizontal projection of the space point in question. Thus by a repeated application of this process to one point of a body after another, the horizontal projection or *plan* of the body can be built up.

The second stage of the problem is to find the heights of the space points whose horizontal projections have now been determined. The heights of the stations whose horizontal projections are o_1 and o_2 are supposed to be known, so that the line o_1o_2 can be graduated, in the manner indicated in the section on contour lines (§ **5**), with heights derived from these. The lines o_1a_1 and o_2a_2 are the horizontal projections of the lines of greatest slope in the two photograms, so that they also can be graduated with heights representing the heights of the corresponding points on the photograms, for the inclination of the plates to the horizontal is known, and the intersection a_1 of o_1a_1 with q_1n_1 produced corresponds to a point on the plate which has the same height as o_1 has on the scale o_1o_2. The point on the first photogram whose projection is s_1 has the same height as the point whose projection is r_1, and the height of r_1 is marked on the scale so that the height of s_1 is known. Find the point g_1 on the scale o_1o_2 which has the same height as s_1, and join s_1g_1 so that s_1g_1 is the projection of a horizontal line in the plane which contains the optical centres of the camera in its two positions and also the space point. Hence if p be the horizontal projection of the space point, and if pg be drawn parallel to s_1g_1 to meet o_1o_2 in g, then the required height of the space point will be the height registered on the scale at g.

58. Discussion of the Problem when certain Data are wanting.—It may happen in practice, that at the time of taking the photograms there were no facilities for obtaining the requisite data as to the true position and aspect of the camera. Thus the only materials on which to work are the actual photograms themselves. The question then arises as to whether these photograms are of any value as a means of obtaining a true metrical representation of the object concerned.

Suppose that two photograms are taken, C_1 and C_2 being the

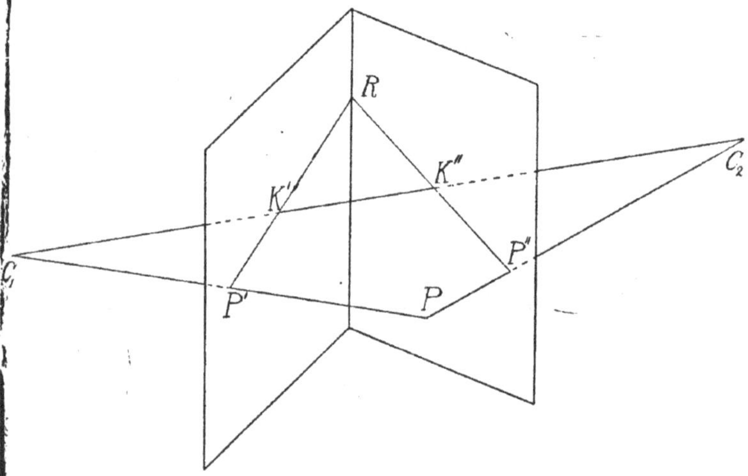

FIG. 42.

(unknown) centres and P' and P'' the representations in the planes of the photograms of some space point P. Then if the C_1C_2 cuts the photograms in K' and K'' respectively, the lines $P'K'$ and $P''K''$ will intersect in some point R on the common *axis* or line of intersection of the planes of the photograms.

The points K' and K'', which are destined to play an important part in the present question, may be called the *Principal Points* * of the photograms. They may always be determined if the projections of four points P_1, P_2, P_3, P_4 which lie in one plane, and those of two other points P_5 and P_6 which do not lie

* Ger. *Kernpunkte*; see Hauck, *Crelle*, 95 (1883), p. 1.

in this plane, are known. For, suppose the straight line C_1P_5 to be drawn, cutting the plane of P_1, P_2, P_3, P_4 in Q_5. The projection of Q_5 on the first plane will be P_5'; suppose that its projection (as yet unknown) on the second plane is Q_5''. Then the two sets of five points P_1', P_2', P_3', P_4', P_5' and P_1'', P_2'', P_3'', P_4'', Q_5'', being the projections of five points which lie in one plane, are in projective correspondence. This correspondence is, however, fully determined since four pairs of corresponding points P_1' and P_2'' . . . P_4' and P_4'' are known, and consequently Q_5'' is determinate and can be constructed by the methods of projective geometry. Similarly, the point Q_6'' which corresponds to P_6' can be constructed. Then since the points C_1, P_5, and Q_5 are collinear, so also are K'', P_5'', and Q_5'' and likewise K'', P_6'', and Q_6''. The required point K'' is the point of intersection of the two straight lines $P_5''Q_5''$ and $P_6''Q_6''$. In like manner the point K' may be constructed. It may be noted that in any photogram which contains representations of architectural structures, it is usually possible to recognise four points which correspond to four space points which lie in one plane.

Now choose any straight line in the first photogram to serve as axis. If P_1' and P_1'', P_2' and P_2'' . . . are pairs of corresponding points in the two photograms, the pencils K' (P_1', P_2', . . .) and K'' (P_1'', P_2'', . . .) will be in projective correspondence, and the range R_1, R_2, . . . in which the first pencil is intersected by the chosen straight line will be projective with the second pencil. This being so, it is always possible so to displace the second photogram that the latter pencil is in perspective with the range, so that the chosen straight line actually becomes the common axis of the two photograms. The angle between the planes of the photograms may be chosen quite arbitrarily and the station points C_1 and C_2 taken at random on the line $K'K''$. The lines C_1P' and C_2P'' where P' and P'' are corresponding points on the two photograms will now intersect in some point P in space, and, as P' and P'' vary together, the point P will trace out a figure in space of which the given photograms are true perspective representations. Since the straight line which was to serve as axis may be chosen in a twofold infinity of ways, and the two station points and the angle between the planes each in a onefold infinity of ways, there is a fivefold infinity of figures in space of which the two given photograms are true perspective images. The knowledge of certain data limits this fivefold arbitrariness to a greater or less extent. Thus, if it is known that in each case the photographic plate was vertical at the moment of exposure, the common axis would have to be taken vertical. If, in addition, the aspects in which the two photograms were taken be known, the angle between the planes is fixed and thus a further degree of constraint is introduced. When the five fundamental data have been either determined or arbitrarily chosen, a solution of the problem may be obtained by the methods outlined in the preceding section.

Spotted something that isn't quite right?

Let us know and we'll fix it for future reprints.
And as a thank you for your diligence, we'll
send you a free Sothis book of your choice.

publisher@sothis.press

This page intentionally left blank

Made in the USA
Monee, IL
05 November 2026

67815673R00059